▶ *"Bro 'Mac' has done it again!* Taken arguments that claim a *factual* basis for the theory of evolution and deny God's existence, then carefully examine them. In my career, I've interviewed the best story-tellers on the planet; I know the truth when I see it."

• Greg Bilbo **(Publisher, Writer, Retired Cop)**
www.bigmacpublishers.com

▶ *"Where will you be 1 million years from today?* If you truly want to know your destiny and sincerely desire to seek the truth, this book is a *must read!"*

• "Rockstar" **(Author "Wild Men, Wild Alaska," Big Game Guide, National Speaker)** www.alaskan-adventures.com

▶ "In his book, 'Faith of an Atheist,' Floyd McElveen brings together, face to face, many of the assertions of modern-day science and the incredible reality of God. Sometimes these clash, as when evolutionists insist that all species on Earth began from a common source, bacteria; other times they harmoniously agree, as when medical scientists describe the incredible design intricacies of the human eye, which itself confirms a supreme designer.

McElveen takes on major arguments of today's scientific community, including the Big Bang Theory, Macro-Evolution, and the age of the Earth, articulating their self-contradictions as well as their acute variance

Big Mac Publishers

with the consistent Word of God, the Bible. He points out that these unproven theories are promoted as fact by largely atheistic groups, revealing the remarkable *faith* of the atheist ... faith that Creation, as we observe it, is simply an accident of nature.

On a foundation of Godly love for the lost, whether atheist, agnostic, or the non-committed, McElveen has effectively written his book to illustrate the need to recognize the fallacies of Godless, humanistic forces at work in our society today, and illuminate the reliability of the Bible and love of God for all. He smoothly transitions his book from its scientific discussion, to evidences of the Bible, to soul-saving evangelism.

Throughout "Faith of an Atheist," Floyd McElveen exhibits his wit and humor, as in his other works, as well as his ability to go right to the heart of the toughest issues that keep millions of people from experiencing the love of God and Jesus Christ and an eternity with Him. The book is uniquely concluded with a wonderful short story by Floyd's wife, Virginia."

•Michael Hockett

Mike is a retired USAF Colonel—43 years and 19 duty assignments. While off-duty, he often served as Local Rep and Area Coordinator for the International Officers' Christian Fellowship, home Bible studies leader, and taught Sunday school in base chapels. Mike and his wife, Betty, married 46 years, have 2 children and 7 grandchildren.

Big Mac Publishers

Faith Of An Atheist

Does God Exist? Did Man Evolve from Apes? Modern Theory of Evolution vs. Creation Debate.

Or

Do the Bible and Science Agree? Is there Real Evidence for the Resurrection of Jesus? What is the Creation vs. Evolution Debate all about and why is it important? If Jesus is God, do Heaven and Hell exist?

Floyd C. McElveen

Big Mac Publishers
Riverside, California

Faith of an Atheist

Author: Floyd C. McElveen
Editor: Greg Bilbo
Proofreader / Cover photo, Train photo © 2009:   Mike Hockett
Cover Illustration / Design: Greg Bilbo

Unless otherwise indicated, Scripture quotations are from:
The Holy Bible, New King James Version © 1984 by Thomas Nelson, Inc.
Other Scripture quotations are from:
The Holy Bible, New International Version (NIV) © 1973, 1984 by International Bible Society, used by permission of Zondervan Publishing House
New American Standard Bible (NASB) © 1960, 1977 by the Lockman Foundation
The Holy Bible, King James Version (KJV)

Library of Congress Control Number: 2009924488
Library of Congress subject headings:
1. Religion and Science
        God – Proof
        Atheism
        Apologetics
        Creation – Study and Teaching
        Evolution – Study and Teaching

BIASC / BASIC Classification Suggestions:
1. REL106000   RELIGION / Religion & Science
2. REL004000   RELIGION / Atheism
3. REL030000   RELIGION / Christian Ministry / Evangelism

ISBN-13: 978-0-9823554-0-4
ISBN-10: 0-9823554-0-8
1.0

Published by Big Mac Publishers
www.bigmacpublishers.com / Riverside, California 92504
Printed and bound in the United States of America

Acknowledgements

Nothing works unless God is in it and the Holy Spirit uses it to His glory. With my son Greg's hard work and editing, this book, "Faith Of An Atheist" and "So Send I You" will soon be on the market.

My whole family, my wife Virginia, son Randy, and daughter Ginger, have been an encouragement and help in my writing career for Christ, and we know many have been saved.

My son, Rocky owns a hunting/fishing lodge in Alaska, and speaks all over America during the few months he spends back in California. He has spoken a number of times for Focus on the Family, and other groups. He is used by churches to speak nationwide, and has had as many as 2,500 men out to a recent meeting, with many decisions for Christ.

He wrote a book, which is very popular, "Wild Men, Wild Alaska," that is published by Thomas Nelson.

In His abundant Love,

Bro. "Mac" (Rev. Floyd C. McElveen)

Faith of an Atheist

Big Mac Publishers

Table of Contents

Faith of an Atheist

Big Mac Publishers

Introduction

As Paul cried out in Rom. 10:1, reflecting the love of God for his fellow Israelites, "Brethren, my heart's desire and prayer to God for Israel is that they might be saved."

So my heart cries out for all unsaved, but particularly for atheists and their evolutionary "faith." As I wrote this book, I sought by God's grace to contrast the "evidence" atheists base their faith on, with the evidence Christians base their faith on. Hang on!

"Trust in the Lord with all thine heart; and lean not unto thine own understanding. In all thy ways acknowledge Him, and He shall direct thy paths." (Prov. 3:5-6)

Only God can make this book useful to His glory, and I pray He will. It is meant for the general public, which seems to be mostly unaware of the titanic battle being waged in our schools, high schools and colleges, in the media and even in some of our churches. For proof of this, see the last two articles in this book.

It is meant for the students and young adults themselves, increasingly bombarded with salvos of unbelief in the guise of evolutionary "fact." It is meant for somnolent Christians sleepwalking their way through

Big Mac Publishers

the minefields of death to their faith and destroying their freedom in Christ.

It is to alert them to the deliberate shattering of the faith of their children, their trust in the Bible and their hope in Christ. This book is not meant to be pejorative, but to be used as a heartbroken rescue mission to all, Christian, unbelievers, atheists, evolutionists and those who have lost their faith, or never had any.

Truly, as Rom. 10:17 says, "So then faith cometh by hearing, and hearing by the word of God."

Beyond all that, it is to reach precious people for the Lord Jesus Christ. As far as I know my heart, it is written in the *love of Christ.*

If Jesus wept over Jerusalem, and He did, God help us to share His burden and weep over the unsaved, as we all once were, whatever tag may be attached to them. God loves them and wants them saved, Jesus died for them, and now, raised from the dead, He pleads with us to love them and reach them. May the Lord Jesus see fit to use this book to turn many to Himself.

Faith of an Atheist

Some atheists believe that there is no God, and that there was nothing in the beginning. Therefore, this type of atheist must believe that NOBODY plus NOTHING equals EVERYTHING! What incredible faith!

In fact, the beleaguered evolutionary atheists believe things by faith even when the facts of his own discipline are arrayed against him. For instance, Dr. Colin Patterson, senior paleontologist at the British Museum of Natural History, with one of the largest collections of fossils in the world, numbering in the millions, admitted that there was not one example of any fossil that could be identified as showing a transition between species. There are no missing links! If evolution were true, and since there are an untold number of "gaps" needing a link, shouldn't there be thousands of these? The fact there are zero *verified* missing links is very telling.

Of course, there are changes within a species, called Micro-Evolution, but no Macro-Evolution exists. God said clearly that reproduction would be "after their kind," and cats are still cats, dogs are still dogs, apes are still apes, and men are still men, from the beginning of Creation. Yet to sustain the theory of evolution, evolutionary scientists, and atheists, must have faith that there are "missing links," and they do.

Or, an atheist subscribes to the "scientific" theory that the created universe always existed. It was therefore never really created; it just evolved. This is totally contrary to the law of entropy, the second law of thermodynamics. As the law of physics confirms, every system proceeding on its own always proceeds in a direction from order to disorder. There has never been a system discovered that has shown otherwise, yet to say that very thing is what happened to our earth and mankind is exhibiting a lot of unfounded faith.

This effectively eliminates the "Big Bang" theory, and really renders a deathblow to the evolutionary theory. As my friend, Dr. Roger Oakland says, "If you take a brand new car, place it in a garage, and leave it there for 100 years without using it, it would eventually deteriorate, *not* become something better." ("'The Bible Key to Understanding The Early Earth,' Dr. G.S. McLean, Roger Oakland, and Larry Mclean, pg. 102.")

In the same book, on pg. 24, Roger Oakland, formerly an atheistic professor teaching evolution in Canada, and his cohorts, say this about an "Early Earth," as opposed to the millions and even billions of years postulated by the sheer speculations of some scientists; "The science of geochronology deals with the subject of determining the age of the earth. At present there are over eighty different methods used in the science of geochronology. Most people are unaware that the majority of these methods result in a *young* age for the earth and not the proposed billions of years as strongly upheld by the evolutionists."

They use only what helps their cause. That is not true unbiased science. Therefore, I conclude that it is more "scientific" to believe in an Early Earth, than the unreliable minority methods that may seem to indicate an earth billions of years old.

On pg. 27-28 of that interesting book, it tells us that the strength of the earth's magnetic field, which had been measured and the data analyzed over the past 130 years, has been getting weaker and weaker each year. The conclusion that is drawn then is that, if we assume that the rate of magnetic decay has been the same in the past as it is today, the strength of the earth's magnetic field would have been equivalent to a magnetic star *only* 10,000 years ago.

If that were true, no life could possibly have existed under those conditions. If the graph were extrapolated only back as far as 30,000 years, then the strength of the earth's magnetic field would have been sufficient to generate temperatures in excess of 5000 degrees Celsius. This temperature is hot enough to melt or vaporize the elements of the earth. According to this method of geochronology, there is evidence to show the earth cannot be as old as suggested by the evolution model.

A friend has pointed out to me that a science researcher/science writer named Robert Felix, has declared that it is now almost a proven fact that magnetic reversals have taken place periodically, with cataclysmic results, "so the assumption is that the rate of magnetic decay has not been the same in the past as it is today." He loaned me a DVD ("Geomagnetic Reversals & Ice Ages," 1998) by Felix, which to me is very unconvincing, but I want to be fair so I recommend that anyone interested get it and decide for themselves.

Felix believes and teaches that rather than global warming, we are now on the verge of a horrific ice age, thus contradicting the astute findings of thousands of scientists who vociferously proclaim the perils of the global warming we are supposedly now entering, a' la Al Gore, with man contributing heavily to this ominous development.

This, if true, undercuts one of the long-held scientific postulations called uniformitarianism—which says that all things have basically continued as they were from the beginning. Thus, the present is supposed to be the key to the past.

Nevertheless, even using the calculations of Robert Felix, the earth would have been an untenable place to live just a few thousand years ago. Frankly, I believe scientists on both sides of the "rate-of-decay" debate for the earth's magnetic field pretty well cancel themselves out, and neither side really fits in with the Biblical model as I understand it, and I am not alone.

I have in my possession a book titled, "In Six Days," edited by PhD John F. Ashton. This book gives the testimony of 50 scientists, all of whom are PhDs. Their fields include: Mechanical Engineering, Biology, Biochemistry, Medical Research, Physical Chemistry, Physics, Mathematical Physics, Organic Chemistry, Mathematics, Botany, Theoretical Chemistry, Medical Physics, Geophysics, Zoology, Nuclear Physics, Meteorology, Medical Research, Orthodontics, Geology, Electrical Engineering, Geological Engineering, Geography, Architectural Engineering, Hydrometallurgy, Meteorology, Forestry Research, Physical Chemistry, Chemistry, Information Science and Agricultural Science.

These scientists *all* believe in the Biblical Six Day Creation, and an Early Earth. According to Ashton, there are a growing number of highly educated critically thinking scientists who have serious doubts about Darwinian evolution; hence, they have *chosen* to believe in the Biblical version of Creation.

The authors of "The Bible Key To Understanding An Early Earth" go on to point out other cogent proofs that the earth is young, and explain what the effects of the earth's decreased magnetic field would have. The Van Allen radiation belts that surround the earth would be impacted. These belts are very important in determining how much cosmic radiation comes in upon the surface of the earth.

Cosmic radiation is an important factor in determining the rate of radioactive Carbon-14 formation. Carbon-14 is a popular method used for dating organic material, *based on the assumption that the amount of radioactive carbon in the earth's atmosphere has always been constant.* As the authors say, "If there has been any fluctuation of the earth's magnetic field in the past, then the accuracy of this method would be highly suspect."

Think about it. This puts evolutionist and atheists in a tough position. If the rate of decay of the magnetic field is constant (as is required for the Carbon-14 dating method to be accurate), and which is used extensively to demonstrate an ancient earth, then there is a huge contradiction. If the rate of magnetic field decay is indeed constant, then the earth *has* to be very young as we demonstrated above. Hey, you can't have it both ways!

I'll say! This discrepancy was shown in Lunar soil dating, listed on pg. 37, with 5 different dates given, varying by billions of years, using 5 different "scientific" methods, Ph207-Ph206, Ph206-U238, Ph207-U235, Ph208-Th232, and Potassium-Argon.

Now let's look at the reliability of Carbon-14 dating. On pg. 45, of the Early Earth book we have been quoting from, we have this information: 1. Living mollusks—that's right, they were alive when they were tested, have been dated by the Carbon-14 procedure and assigned an age of *2300* years old; 2. A castle known to be 787 years old was reported by the Carbon-14 method to be 7370 years old; 3. Freshly killed seals were dated by the Carbon-14 method as being

1300 years old. Mummified seals that had been dead for 30 years tested at 4600 years old. In each case, the book lists the scientific journals and other sources from which this information was obtained.

Several of the most graphic examples of erroneous dating are reported on pg. 38 concerning lava rocks formed in Hawaii in 1800 and 1801, dated as being 160 million to a billion years old. The book continues with a report from "Science" magazine, volume 162, October 11, 1968, about volcanic rocks, known to be less than 200 years old, yet were dated by a radiometric dating method at ages of 12 to 21 millions of years old. Naturally, these same unreliable dating methods are used by evolutionists to prove the age of the earth, and presented as *infallible fact.* Many of the same people laugh in derision at our believing the Bible to be the infallible Word of God, yet present their findings with imperturbable confidence. What faith…or folly!

Concerning Roger Oakland, he has written several fine books, one of them in conjunction with T.A. McMahon, "Understand the Times." Roger has put out a terrific video on Creation, "A Question of Origins."

His complete change of mind from evolutionary tentacles, to freedom in Christ, came about as a direct result of a real conversion to the Lord Jesus Christ, and further serious study of the whole issue. Roger is no insipid neophyte as a scientist. He spoke by *their* request, to 300 of Russia's top scientists at one time.

As far as America is concerned, the atheist is free to be an atheist, Muslims are free to be Muslims, Mormons are free to be Mormons, Baptist are free to be Baptists, Evangelicals are free to be Evangelicals, etc., in this great country of ours. That is not the point. We need to be especially kind and tender, speaking the truth in love, to those we seek to win to Christ, but sometimes each group abandons reason and seeks to demonize those who differ. We are admonished in Jude to "...contend for the faith" but without being contentious.

We are not to be *so* tolerant that we abandon Biblical truth. Dr. Josh McDowell graphically depicts how fallacious this modern "accept everything and everyone on an equal basis, regardless of how they may violate not only the Bible, but common sense." Suppose you were talking to someone about Jesus Christ being the Son of God and he responds after pondering this Biblical truth, "Well, I don't know. Ronald McDonald is a wonderful person, and has done many good works. I believe that he is the Son of God!"

While this individual has the freedom of speech to say such an outlandish thing, modern society has so degenerated from truth, that he must be accorded equal consideration and his belief endorsed as important and on the same level as the Biblical teaching based on an incredible amount of evidence that Jesus Christ was and is the Son of God. This is an endorsement of nonsense at best, and sheer lunacy at worse, and unfortunately America is moving at mach speed in that direction.

If this continues, how is one to know the truth about anything? All history will be continually redacted and revised until it becomes either incomprehensible, or fills some group's agenda, or totally abandons the truth. To some extent, this is already taking place in our schools, and World and American History. It is a threat in religion, science, medicine, and many other disciplines.

After all, the modern, sophisticated, supercilious, science sophisticates say sanctimoniously, "As we all know, there are no absolutes." Of course, that is a self-defeating statement, like saying, "Absolutely there are no *absolutes!*"

Even *science!* Science cannot agree on many simple things. After reading 10 or so studies over the years on coffee consumption—some of the studies were major, involving thousands of people—there are still no consistent conclusions. We have all read contradicting "Scientific findings." Such as, "Coffee is bad for you; coffee is good for you; it hurts your heart; no, it should be listed as a health food, it is that good for you," and other statements that run the spectrum from bad to good. I still don't really know if coffee is *good for me or not*. Science can't decide, except momentarily, as reflected in the latest coffee study.

And of course, although we have been to the moon, we still can't unalterably cure the common cold.

Yet *Science* is the only God some people, especially atheists, have. "Faith" is at work again!

Perhaps this is the time for a caveat in relation to real science. I love and appreciate true science, not the inane scientific speculations some infer to science. We all have benefited immensely from the hard, meticulous work and research of dedicated scientists, in technology, in overcoming certain diseases and in the quality of life.

I owe my own life to the intervention of the grace of God through science, as I had a heart transplant years ago, May 9, '94. Without real science, I would not be here. I don't believe that real science ever collides with the Bible. I do know many Christian scientists, John Morris, Ken Ham, Roger Oakland, John Whitcomb, the late Henry Morris, etc., and many other scientists who are not Christians, that nevertheless, are still hard-working, dedicated men, really seeking truth and advancing human progress on many fronts.

Too many others, especially atheists and skeptics, flee to science as their palladium of defense against the Bible, and against Christians, often making mendacious statements scientists themselves would not make. One of many mistakes they make is saying that no "respectable" scientist could ever believe the Bible, the Creation story, the supernatural, and the resurrection of Jesus Christ. The previously identified book published in 2001, "In Six Days," edited by John Ashton, gives the testimony of 50 scientists who *choose* to believe in Creation. Moreover, each of these scientists is *very respectable.*

Yet, yesterday's scientific "facts" are often contradicted by today's science. The textbooks are obsolete in a few years, or subject to radical revision—not so, the immutable, infallible Bible!

What incredible faith an individual must have to trust that science is true and the Bible, Heaven and Hell, and the risen Lord Jesus Christ is a *myth.* What faith atheist must have to trust their looming eternity, and the answers to why they are here; where they came from; and where they are going when they die, to men like themselves, flawed, and however brilliant, knowing only a modicum about this world and much less about the one to come.

As I stated in one of my books, "It is reported that Einstein once gave the cogent illustration that if all the wisdom, of all the men who ever lived, was amassed together, we would still know *less* than 1% of all there is to know" (Straight Thinking About Crooked Ideas, pg. 28) Yet with this infinitesimal bit of wisdom, of which the smartest man would only know a much smaller infinitesimal percentage, we dare to trust ourselves, our knowledge, rather than an Omniscient, Sovereign God, the God of the Bible revealed in Jesus Christ.

Whatever happened to *truth?* Does truth matter? You had better believe it; we are betting our very soul it does! At this point, I highly recommend to doubters that they read Bill Wilson's compilation of Josh McDowell's book, "A Ready Defense." Especially the chapter beginning on pg. 241, "The Trilemma—Lord, Liar, or Lunatic?" I recommend as well as a similar book by Dr. John Ankerberg, from the John Ankerberg Show with guest John Warwick Montgomery, "Jesus Christ: Was He A Liar, A Lunatic, A Legend, or God?" This evidence is presented with great clarity and power. Do you really want to know the truth? Read them both.

When dealing with many of our atheistic friends, as well as in our reading of some of their writings, such as "The Age of Reason," by Tom Paine, and others more recent, we as Bible believers, are obviously viewed contemptuously and considered to be naïve and intellectually inferior.

In incredulous disbelief, some skeptics say, "You mean that you believe a *Snake* talked? You believe God created the World in 6 days? You believe God actually had a huge fish swallow Jonah, and kept him alive in the belly of the fish 3 days? You believe a loving God would send His children to Hell to burn forever, and so on." (God *never* says He will send His children to Hell, but only those who refuse to become His Children by being born-again into His family when they receive the Lord Jesus Christ, and His forgiveness through His shed blood.) (John 1:12)

It is no wonder that God says, in Psalms 14:1a, "The fool hath said in his heart, 'There is no God.'"

Obviously, the fool has said as well in his heart, "No," to God! That really is how the unbelief started; all the rest is a futile attempt to bolster this rebellion. Imagine these highly intelligent men, for the most part, actually being *fools*.

Somewhere, they began to question what God said. My dear Greek professor, now deceased, Dr. Milliken, who had been a missionary in Canada, told us that when someone told him they were an atheist, and didn't believe in God, he then asked them what sin they were committing that they did not want to give up?

Sometimes, atheists come from fundamental Christian homes and rebel against their Christian upbringing and other doctrines of Jesus. "No loving God would ever send anyone to Hell," is a common assertion among them. Or their fundamentalist parents were too strict, or too legalistic, or they were hypocrites and did not live what they professed, etc. Many of those, who are not yet full-fledged atheists, invent a Jesus, and a God of their own, while sadly fooling themselves into thinking that they are "Christians." What they have done is create God in man's image. They are en route to the empty life of atheism and an eternity without the God I know and love.

On April 11, '08, I received a call from Joel Groat, an internationally known speaker on cults, and especially Mormonism, now interim director of the Institute of Religious Research, located in Grand Rapids,

Michigan. He told me they were all ready to reprint my "Evidence," book, which God has chosen to use. I mentioned to him that I was working on a book, the working title of which is, "The Faith of an Atheist," and he quoted me Romans 1:18-22, which states:

"For the wrath of God is revealed from heaven against all ungodliness and unrighteousness of men, who hold the truth in unrighteousness; Because that which may be known of God is manifest in them; for God hath shewed it unto them. For the invisible things of Him from the creation of the world are clearly seen, being understood by the things that are made, even His eternal power and Godhead; so that they are without excuse; Because that, when they knew God, they glorified Him not as God, neither were thankful; but became vain in their imaginations, and their foolish heart was darkened. Professing themselves to be wise, they became fools..."

God forbids that we should call any man a fool, but God knows the heart, the head, the motives, etc. He consistently calls those who deny Him and His Creation, fools. Incidentally, many evolutionists are not atheists, and some of them are professing Christians, but it is a tenuous marriage at best. It may be that some atheists are not evolutionists, but insofar as my experience goes, evolution and atheism go together like a hand in a glove. Where else can the atheist go, once he has abandoned God, Creation and the Bible?

To any honest seeker who may read this, please consider this; one cannot go to a banker, and set all the terms about getting a loan. We cannot go to God and tell Him that He must do thus and so for you to believe in Him. And we will not ever be successful in finding Him, if we do not heed what He said.

Knowing men and our colossal egotism, God explains in 1 Cor. 1:18-25, "For the preaching of the cross is *to them that perish* foolishness; but unto us which are saved it is the power of God. For it is written, I will destroy the wisdom of the wise, and will bring to nothing the understanding of the prudent. Where is the wise? Where is the scribe? Where is the disputer of this world? Hath not God made foolish the *wisdom of this world?* For after that in the wisdom of God, the world by wisdom knew not God, it pleased God by the foolishness of preaching to save them that believe. For the Jews require a sign, and the Greeks seek after wisdom. But we preach Christ crucified, unto the Jews a stumbling-block, and unto the Greeks foolishness; But unto them which are called, both Jews and Greeks, Christ the power of God, and the wisdom of God. Because the foolishness of God is wiser than men; and the weakness of God is stronger than men." As God explains in v28-29, in the same passage, this is so "That no flesh should glory in His presence."

God said explicitly and absolutely, that no man will ever find Him by wisdom. That door is *closed.* He will reason with a seeking sinner. He said in an invitation in Isaiah 1:18, "Come now, and let us reason together, saith the Lord; though your sins be as scarlet, they shall be as white as snow, though they be red like crimson, they shall be as wool."

God gives a tremendous, overwhelming amount of evidence for His existence from nature, from the conscience of man (See Rom. Ch. 1-3), but most especially from a plethora of irrefutable evidence in the Bible. That includes hundreds of precisely fulfilled prophecies, given scores, hundreds and thousands of years in advance. These include the miracle birth of Jesus Christ, the incredible return of the Jews to their homeland, the resurrection of the Lord Jesus Christ and the changed lives and immutable peace of those who truly trust Him.

In 1 Cor. 2:14-16, God sheds some light on why the natural, unsaved man *cannot* understand the Bible. This is true even if he graduated from the most prestigious Seminary, and garnered a string of PhDs. "But the natural man receiveth not the things of the Spirit of God; for they are foolishness unto him; neither can he know them, because they are spiritually discerned. But he that is spiritual (a born-again Christian who has the Spirit of God dwelling in him) judgeth all things, yet he himself is judged of no man. For who hath known the mind of the Lord, that he may instruct Him? But we (Christians) have the mind of Christ."

No man *ever* found God by his wisdom, not Plato, not Aristotle, not Socrates, not Einstein, not the Mensa geniuses, not astute college professors, not scientists, *no one.* That door is *closed.* God said so. So why, atheists, skeptics and some unsaved are so many of you determined to find God by wisdom, when He said you could not, and no man ever has or ever will. Is that an intelligent decision? Then, when you do not find Him by your wisdom, why do you declare pontifically that there is *no God?*

In a book I wrote years ago titled "Straight Thinking About Crooked Ideas," I gave this illustration, pg. 43-44:

"Suppose you are driving down an unfamiliar highway en route to a town called Bunkersville. You come to a 'Y' in the road, with both roads continuing in the general direction of Bunkersville. One road is marked 'Closed,' and the other road is marked 'Bunkersville.' Which road would you take?

The sensible action would be to take the road marked 'Bunkersville.'

But suppose for some reason—or for no particular reason—you doubt the signs. You hesitate, uncertain what to do, and then you notice something. Driving back out of the road marked 'Closed,' you see a continual stream of traffic, cars battered and drivers hag-

gard, bitter, and disillusioned. Some of them now go down the Bunkersville road, but others shake their heads sadly and say that there simply is no way to Bunkersville. Though you point out the sign that clearly indicates the way, they refuse to take the right road. Some drivers even contend that there is no Bunkersville, because they could not reach it on the road marked 'Closed!'

Meanwhile, some drivers passing by tell you with great joy and assurance that they have already tried the road marked 'Bunkersville,' and that it really does lead there."

What would be your judgment about the only intelligent action for you to take? How can you, atheists and some other unbelievers, dare laugh at, taunt, and mock Christians and think yourselves superior to them in your supposed great intellectual attainments and thinking ability, when you have made the most foolish and eternal blunder any human being could possibly make? You have concluded that there is no God! You have attempted to find Him possibly, or deny Him, by man's wisdom, a proven impossibility, forbidden by God, just like *Bunkersville!*

God's road sign, the Bible, says the only door that leads to Him is marked faith. He says man's wisdom never leads to Him, That door is marked *Closed.* (1 Cor. 1:21) Yet He will give any willing mind more than sufficient evidence to make faith credible. However, He will not give such overwhelming proof that no faith is required, because He has established faith as the route to God.

Deuteronomy 20:29 gives the key to finding God, "But if from thence thou shalt seek the Lord thy God, thou shalt find Him, if thou seek Him with all thy heart and with all thy soul."

Once found, based on 1 Cor. 15:1-4, encompassing the Gospel, Eph. 2:8-9 becomes an immediate responsibility. "For by grace are ye saved through faith, and that not of yourselves; it is the gift of God; not of works, lest any man should boast."

You say you need evidence? Most of us (who know Jesus and have studied the Bible), are stunned when a callow youth, or a compromising or uninformed Christian, or a PhD atheist, asserts with irrefragable authority, that there is no real, *solid* evidence for Christ or the Bible.

While we try to be gentle and nice and we want to speak the truth in love, we know that this is a statement of colossal ignorance. If you want more evidence…

Consider the Human Eye.

For the following information about the design of the human eye, I am grateful to the Southern Eye Center of Hattiesburg, Mississippi, and the help of Chris Crawford employed there, for much of the following, which is used by permission.

The human eye is made up of the Iris, the Cornea, the Lens, the Pupil, etc. What astonishes me is that it contains 120 million *rods*, and 6 to 7 million *cones*. There are 1000 billion cells in the human body, and all of these came from one cell, with its almost incomprehensibly complex DNA. Concerning the eye, the retina contains two types of photoreceptors, rods and cones. The rods are not sensitive to color. The cones provide the eye's color sensitivity and they are much more concentrated in the central yellow spot known as the macula. In the center of that region is the "fovea centralis," a 0.3 mm diameter rod-free area with very thin, densely packed cones.

The experimental evidence suggests that among the cones there are three different types of color perception. Response curves for the three types of cones have been determined. Since the perception of color depends on the firing of these three types of nerve cells, it follows that visible color can be mapped in terms of three numbers called tristimulus values. Color perception has been successfully modeled in terms of tristimulus values and mapped on the CIE chromaticity diagram.

Measured density curves for the rods and cones on the retina show an enormous density of cones in the fovea centralis. To them is attributed both color vision and the highest visual acuity. Visual examination of small detail involves focusing light from that detail onto the fovea centralis. On the other hand, the rods are absent from the fovea. At a few degrees away from it their density rises to a high value and spreads over a large area of the retina. These rods are responsible for night vision, our most sensitive motion detection, and our peripheral vision.

Current understanding is that the 6 to 7 million cones can be divided into "red" cones (64%), "green" cones (32%), and "blue" cones (2%), based on measured response curves.

Since the cones are less sensitive to light than the rods, as shown in a day-night comparison, the daylight vision (cone vision) adapts much more rapidly to changing light levels, adjusting to a change like coming indoors (automatically, I might add) out of sunlight in a few seconds.

Like all neurons, the cones fire to produce an electrical impulse on the nerve fiber and then must reset to fire again. The light adaption is thought to occur by adjusting this reset time.

The cones are responsible for all high-resolution vision. The eye moves continually to keep the light from the object of interest falling on the fovea centralis where the bulk of the cones reside.

The rods are the most numerous of the photoreceptors, some 120 million of them, and they are more sensitive than the cones. However, they are not sensitive to color. They are responsible for our dark-adapted or scotopic vision. The rods are incredibly efficient photoreceptors. More than one thousand times as sensitive as the cones, they can reportedly be triggered by individual photons, under optimal conditions. The optimum dark-adapted vision is obtained only after a considerable period of darkness, about 30 minutes or more, because the rod adaption process is much slower than that of the cones. While the visual acuity or visual resolution is much better with the cones, the rods are better motion sensors. Since the rods predominate in the peripheral vision, that peripheral vision is more "light sensitive," enabling you to see dimmer objects in your peripheral vision.

The rods employ a sensitive photopigment called rhodopsin.

Does this sound complicated? Any scientist could tell you that I have *barely* touched on the incredible function and complexity of the human eye. Because of the "irreducible complexity" illustrated elsewhere in this book, by the mousetrap illustration, the eye could not have evolved. It was designed and functioning in infinite detail from the very beginning of its existence.

Now read how desperate and even foolish some of the top evolutionary scientists have become, even as some of them and their atheistic allies disavow the Christian God, and become more and more aggressive in our culture. I am using this excerpt by permission, from "The Berean Call," edited by Dave Hunt and Tom McMahon, January edition, 2009; it is long, but please hang in there.

Questions: In Christopher Hutchens's "God is Not Great," he says that the magnificent, irreducible complexity of the human eye is not evidence for a Creator, but cites "the ineptitude of its design" as proof for evolution. He quotes Dr. Michael Shermer, who claims, "a simple eyespot with a handful of light-sensitive cells...developed into a recess eyespot...then into a pinhole camera eye...then into a pinhole lens...then into a complex eye." Shermer goes on to say, "The anatomy of the human eye, in fact, shows anything but 'intelligence' in its design. It is both upside down and backwards, requiring photons of light to travel through the cornea, lens, aqueous fluid, blood vessels, ganglion cells, amacrine cells, horizontal cells, and bipolar cells, before they reach the light sensitive rods and cones that transducer the light signal into neural impulses...which are then sent to the visual cortex at the back of the brain..."

Hutchens says, "It is because we evolved from sightless bacteria, now found to share our DNA, that we are so myopic...we must never forget Charles Darwin's injunction that even the most highly evolved of us will continue to carry 'the indelible stamp of their lowly origin.'"

Hutchens asks "For optimal vision, why would an intelligent designer have built an eye upside down and backwards?"

Response: I have a question first for Hutchens. Can he prove that we "evolved" from sightless bacteria? It is true that we all share the same DNA alphabet, even with carrots and garden slugs; but human DNA is far more complex. Nor is DNA all that makes us human and separates us from all lower creatures. What part of the DNA spells out appreciation for poetry, the ability to compose an opera, or to write like Shakespeare, or Dickens? Where does the DNA spell out the genius to define the mathematics to engineer the construction of a high-rise building or to design the space capsule that landed on the moon? None of these abilities came from DNA, nor even from the brain, but from the *nonphysical* mind.

Hutchens is determined to support his atheism at any cost, and that makes him so eager to accept anything that seems to do so that he is blind to the many facts to the contrary. The truth is that Shermer, upon whom Hutchen relies, has the facts twisted. He recites the standard theory of evolutionists concerning the origin of the eye from "a simple eyespot with a handful of light sensitive cells...then into a pinhole camera eye...then into a pinhole lens...then into a complex eye."

Evolutionists all repeat this same recital as though it has been established by fossils, but that is far from the case. It doesn't take a genius to realize that this is pure speculation. A child could ask simple questions that neither Shermer nor Hutchens could answer. What is an eyespot? How did it develop? Many cells on our body are "light-sensitive" but none of them will turn into an eye…why this one?

How and why did it develop a "recessed eyespot?" How did it make the huge leaps from "eyespot" to a "pinhole camera eye" then into a "pinhole lens"—and how could it have been called a "camera" before it had a "lens?" At what point did these partial developments begin to benefit the organism enough to aid in its survival? How did they avoid being wiped out by natural selection before they became part of a functioning whole?

As for the eye being badly designed, ophthalmic scientists have denounced this idea. For example, Dr. George Marshall and Sir Jules Thorn Lecturer in Ophthalmic Science University of Glasgow declare, "The belief that the eye is wired backwards comes from a *lack of knowledge* of eye function and anatomy. The nerves could *not* go behind the eye, because that space is reserved for the choroids, which provides the rich blood supply needed for the very metabolically active retinal pigment epithelium (RPE). This is necessary to regenerate the photoreceptors, and to absorb excess heat. So…the nerves (must) go in front instead."

"Inverted wiring is necessary for vertebrate eyes to work…the direct *opposite* of what evolutionists claim would be the 'correct' wiring. In fact, the evolutionist claim is actually undercut by their own assessment of squid eyes, which despite being 'wired correctly,' don't see as well as vertebrate eyes…"

"Interestingly, anyone with excellent eyesight is said to have 'eyes like a hawk,' which are 'backwardly wired,' not 'eyes like a squid.' The excellent sight provided by these allegedly 'wrongly wired' eyes makes (evolutionists') objections absurd... (The) claim that the nerves obstruct the light has been falsified by very new research by scientists at Leipzig University..."

"Not only is the inverted wiring of our eyes a good design, necessary for proper functioning (but) it is coordinated with an ingenious fibre optic plate. Therefore, the vertebrate eye has the advantage of a rich blood supply behind the receptors without the disadvantage of nerves blocking out light. Such fine coordination of parts makes sense with a Master Coordinator, while it's a puzzle for evolutionists." (This ends the Berean Call article.)

The evolutionists, and their sometimes atheistic associates, are becoming aggressive and intimidating in trying to get the field all to themselves in our schools, colleges and the media; and trying to erase Christianity from our culture, or at least keep its expression walled up in our churches. For more proof of this, see the last two articles in this book.

Yet, consider this. If almost any evolutionists or atheists found a super-sophisticated camera on the beach, with its multi-faceted functions, not one of them would be foolish enough or stupid enough to wonder how it made itself, how it "evolved" or how it developed without outside intelligence. They would immediately know that someone made it; it could not make itself.

Yet, because of an almost insane prejudice against Intelligent Design, which they know opens the way to a Creator, and on to the God of the Bible, the Lord Jesus Christ, and to the contemplation of Heaven and Hell, and their rebellion and responsibility before God, they will distort and ignore facts, in the *name* of science. The eye, the human body, the cell, DNA, obviously *did not evolve.*

To claim that the very best camera must have been made by an intelligent designer, and yet that a far more sophisticated, infinitely more complex, and incredibly better functioning entity, the eye, "evolved," is asinine.

Truly, in Rom. 1:22, God says, "Professing themselves to be wise, they became fools."

While I especially seek to protect our youth and adults as well from these evolutionists, and/or atheistic depredations, my heart actually goes out to them. They are dear people too, though they have embraced a soul-damning delusion. I pray that they may open their eyes to the Lord Jesus Christ, and be saved. Some have. May many more read the Bible and seriously consider the Lord Jesus Christ, who died for them, and weeps for them.

Eternity is so long, no one can afford to miss the love of Jesus for their soul.

The Biblical Evidence

There is *so much evidence*, it is difficult to know where to start, and Christian scholars have written thousands, perhaps millions, of pages chock full of evidence, based primarily on the evidence in God's Word as well as their own corroborating experience.

The real problem, however, is well stated by lawyer/author Irwin W. Linton, in his book, "A Lawyer Examines the Bible." After a thorough examination of the Bible, testing it by a lawyer's inculcated, trained scientific analysis, and facing its ineluctable truth, Mr. Linton received Christ. He became a spokesperson for Jesus, doing much traveling and speaking for 25 years or more. The problem is this. It was his assertion that he never, in a multitude of meetings, met *one* atheist, skeptic, agnostic, infidel or unbeliever that had read *even one book* of Christian apologetics.

Unfortunately, from speaking in hundreds of Churches, and witnessing to perhaps thousands of individuals, I have to echo his conclusion. According to the opinion of many atheists, unbelieving college professors, and skeptics in general, Christians believe what they believe purely by "faith" which skeptics may excoriate as being narrow-minded and obscurantist. Yet many of us have read the arguments of atheists, evolutionists and assaults against the Bible. Who is the true obscurantist, when we have read both sides, and they have not? Who is truly seeking the truth?

Look, for instance, at a partial list of prophecies about Jesus, many of them given hundreds of years before He came to earth.

Josh McDowell records in his excellent book, with co-author Bob Hostetler, about the last day or so surrounding the crucifixion. There are more than 300 prophecies concerning the Lord Jesus Christ, yet Josh only deals with 29. ("Beyond Belief to Conviction," pg. 65-67)

1. He will be betrayed by a friend. (Psalm 41:9, Mt. 26:49)

2. The price of His betrayal will be thirty pieces of silver. (Zechariah 11:12, Mt. 26:15)

3. His betrayal fee will be cast to the floor of the temple. (Zechariah 11:13, Mt. 27:5)

4. His betrayal money will be used to buy the potter's field. (Zechariah 11:13, Mt. 27:7)

5. He will be forsaken and deserted by His disciples. (Zechariah 13:7, Mark 14:50)

6. He will be accused by false witnesses. (Psalm 35:11, Mt. 26, 59-60)

7. He will be silent before His accusers. (Isaiah 53:7, Mt. 27:12)

8. He will be wounded and bruised. (Isaiah 53:5, Mt. 27:12)

9. He will be hated without a cause. (Psalm 69:4, John 15:25)

10. He will be struck and spit on. (Isaiah 50:6, Mt. 26:67)

11. He will be mocked, ridiculed and rejected. (Isaiah 53:3, Mt. 27:27-31, and John 7:5, 48)

12. He will collapse from weakness. (Psalm 109:24-25, Luke 23:26)

13. He will be taunted with specific words. (Psalm 22:6-8, Mt. 27:39-43)

14. People will shake their heads at Him. (Psalm 109:25, Mt. 27:39)

15. People will stare at Him. (Psalm 22:17, Luke 23:35)

16. He will be executed among 'sinners.' (Isaiah 53:12, Mt. 27-38)

17. His hands and feet will be pierced. (Psalm 22:16, Luke 23:33)

18. He will pray for His persecutors. (Isaiah 53:12, Luke 23:34)

19. His friends and family will stand afar off and watch. (Psalm 38:11, Luke 23:49)

20. His garments will be divided and won by the casting of lots. (Psalm 22:18, John 19:23-24)

21. He will thirst. (Psalm 69:21, John 19:28)

22. He will be given gall and vinegar. (Psalm 69:21, Mt. 27:34)

23. He will commit Himself to God. (Psalm 31:5, Luke 23:46)

24. His bones will be left unbroken. (Psalm 34:20, John 19:33)

25. His heart will rupture. (Psalm 22:14, John 19:34)

26. His side will be pierced. (Zechariah 12:10, John 19:34)

27. Darkness will come over the land at midday. (Amos 8:9, Mt. 27:45)

28. He will be buried in a rich man's tomb. (Isaiah 53:9, Mt. 27:57-60)

29. He will die 483 years after the declaration of Artaxerxes to rebuild the temple in 444 B.C. (Daniel 9:24)

As a final testimony, on the third day after His death, He will be raised from the dead (Psalm 16:10, Acts 2:31), ascend to heaven (Psalm 68:18, Acts 1:9), and be seated at the right hand of God in full majesty and authority. (Psalm 110:1, Hebrews 1:3)

What extraordinary lengths God went to in order to help people identify and recognize His only begotten Son. Jesus fulfilled about sixty major Old Testament prophecies (with about 270 additional ramifications)—all of which were made more than 400 years before His birth. This makes a compelling case for the Deity of Christ.

I'll say it does! It really breaks my heart that there will be those in the final Lake of Fire, moaning forever, "Oh my God, what a *fool* I have been." I weep for them.

Lee Strobel, the one-time skeptic and superb investigative reporter, now converted to Christ, wrote a scintillatingly good book, "The Case for the Real Jesus." He says, on pg. 274-275, in view of the prophecies, which he calls identifying fingerprints for the real Messiah, "Against astronomical odds—by one estimate, one chance in a trillion, trillion, trillion, trillion, trillion, trillion, trillion, trillion, trillion, trillion, trillion, trillion, trillion—Jesus, and only Jesus throughout history, matched this prophetic fingerprint. This confirms Jesus' identity to an unassailable degree of certainty."

Actually, the shoe is squarely on the other foot. There is no chance whatsoever that Jesus Christ was not God in human flesh, dying for us on the cross, shedding His blood for us, and rising again from the dead to love us to Himself *forever*.

Compare this with the evolutionary statement I have been seeing on TV a lot lately, "It took 540 million years for the eye to evolve." Do you believe *that,* my skeptic, atheistic, or infidel friend?

In my book, "Evidence That You Never Knew Existed," now in a number of languages, with 360,000 in Russia alone (although under several different titles), plus some in Simplified and Mandarin Chinese, Ko-

~ 33 ~
Big Mac Publishers

rean, Romanian, and on the internet in Spanish, I gave several illustrations which fit here.

"Evidence You Never Knew Existed," by Floyd C. McElveen, pg. 33-36, "Weighing The Evidence."

We have briefly stated some of the compelling evidence for the Bible and for Jesus Christ. Can we identify Jesus Christ beyond the shadow of a doubt as being who He claimed to be? Please consider the evidence one more time. God will not force you to believe. You must make that final choice. But to fail to act on that evidence will haunt you for a billion years in Hell, and that breaks my heart for you. (Now consider this illustration regarding identifying a person in the setting of Philadelphia, Pa.)

When I was a sailor, I had several "blind dates." Suppose for this particular blind date, I agree to meet her at the Greyhound Bus Station, 464 Liberty Street tonight at 8:00 p.m. She tells me she is one-legged, and wears a wooden peg leg, which she has painted florescent yellow with a blinking red light built in it to keep folks from stumbling over it. She has a matching florescent yellow patch over her right eye that she lost in the same accident in which she lost her leg. In addition, she is missing the little finger on her right hand. She will wear a pink stocking on her good leg and a maroon and white saddle oxford on her one good foot. She will wear a green hat and carry a purple purse. She says she is five feet tall and weighs about 200 pounds.

Do you honestly think I would have any trouble identifying the right girl at the Greyhound Bus Station at 8:00 p.m.?

(Now apply this information to the identifying of Jesus Christ).

Remember, the time of Jesus Christ's birth was foretold centuries in advance. (Daniel 9:24-26) This does away with the otherwise possible objection that is sometimes made that one could make up a prophecy and something would eventually occur which might be labeled as the fulfillment of that prophecy. The time element destroys this objection.

I gave only 13 or 14 identification marks concerning my blind date. However, the chances are millions to one against there being another girl with these *same* identifying characteristics in that particular bus station at 8:00 p.m.

God gave 333 marks of identification for Jesus, concerning His birth, death and resurrection. Each mark of identification was fulfilled perfectly in Jesus, so there could be no doubt about identifying Him when He came, and verifying His identity now that He has come. Remember, everything Christ said came true. He said Hell was real and forever just as Heaven is. A million years from now, you will be somewhere. Will it be in Heaven or Hell?

Now let us consider the second illustration from the book.

Every prophecy of the Bible is always accurately, literally fulfilled, and only when symbols or figures of speech make *absolutely no literal sense*, should anything but a literal interpretation of the Bible be sought. The prophecies concerning Jesus clearly meet this test.

To get the full impact of this, suppose some prognosticator predicted 100 things that would happen to you in the coming year. These predictions are very detailed. The first prediction is that you will stub your toe on a chair leg on January first, at 2:35 a.m. You will fall on a glass on the kitchen table, which will shatter and cut a U-shaped wound in your chin. This jagged wound will require 13 stitches. It will be sewn up by a new doctor in town named McGuire, your doctor being unavailable at the time. To your chagrin and amazement, when January first comes, this is *exactly* what happens, right down to the smallest detail.

Then, throughout the year, 99 of these prophecies are literally, actually, perfectly fulfilled—in every detail. Ninety-nine of the 100 are fulfilled, with one more to go. This last prophecy is to happen on the last day of the year. It declares that if you drive down to Fifth and Main at 5:00 p.m. on that day, you will be in a fiery car crash that will leave you blind, crippled and badly burned. You will be in excruciating pain. You will be hospitalized for six months and then die.

Tell me, would you deliberately drive down to Fifth and Main at 5:00 p.m. on the last day of the year, if you had a choice? Would you consider it a safe, sane, intelligent risk when 99 prophecies have come true without a failure, which would be virtually impossible mathematically, by chance alone?

Let's just round the Bible prophecies off to 100 (out of hundreds), 99 of which have already been fulfilled. The one-hundredth prophecy is this. If you ignore or refuse to accept Jesus Christ as your personal Lord and Savior, you will die without hope and spend eternity in the Lake of Fire spoken of in Revelation chapter 20 as the destiny of the lost. A billion years from now, your agony, despair and lostness from His love will have barely begun. Since all the other Bible prophecies have come true in literal and absolute fact, would it be intelligent to gamble your destiny forever that this final prophecy will not also come true?

Atheist friend (and all unbelievers), many of you pride yourself on being more intelligent than we Christians are. Maybe you are. Certainly some I have dealt with are much more intelligent than I am.

God does not decide where we are going to spend eternity on how intelligent we are, but whether we depend on Jesus Christ alone for His Salvation.

But part of the sheer desperation, agony and torture of Hell for you is that you *ignored* the evidence, and made the most unintelligent decision any human being could ever make. As these two simple but clear illustrations show, the evidence for Christ is totally overwhelming and inescapable. And we have not even dealt with the most powerful evidence, the resurrection of Jesus Christ.

On April 10, '08, The Baptist Record of Mississippi recorded a new documentary film on Evolution, by Ben Stein. The film is titled "Expelled; No Intelligence Allowed." It opened April 18. Stein feels the proponents of evolution have emasculated freedom of speech in their opposition to Intelligent Design/Creation.

He interviews Holocaust survivors and others, and shows a very strong connection between evolution's survival of the fittest and Hitler's savage attack, on not only the disenfranchised Jews, but also the weak and suffering. The Nazis thought they were carrying out the Darwinian ideas, or at least they so rationalized.

The film included interviews with leading atheist and author, Richard Dawkins, and atheist, evolutionist biologist PZ Myers. It also included Eugene Scott, executive director of the National Center for Science Education. The film depicts the fact that some prominent scientists (professors who dared to *differ* with the evolutionary concept), *lost* their jobs or couldn't get tenure. That is more like survival of the *conformist.*

Now consider the real evidence a loving God has given to reach lost and confused humanity, because He loves us.

The following article delves deeper into aspects of Nature that clearly demonstrate intelligent design.

God's Accuracy

How awesome to think about the way our Creator God planned everything so carefully and perfectly, everything with a purpose. As His highest creation, "we are fearfully and wonderfully made."

God's accuracy may be observed in the hatching of eggs. For example, the eggs of the potato bug hatch in 7 days, those of the canary in 14 days and those of the barnyard hen in 21 days. The eggs of ducks and geese hatch in 28 days; the eggs of the parrot and the ostrich hatch in 42 days. (Notice, they are all divisible by seven.)

God's wisdom is clearly seen in the making of an elephant. The four legs of this great beast all bend forward in the same direction. *No other Quadruped* is so made. Did that happen by chance? Hardly! God planned that this animal would have a huge body—too large to live on two legs. For this reason, He gave it four fulcrums so that it can rise from the ground easily. The horse rises from the ground on its two front legs first. A cow rises from the ground with its two hind legs first. How wise the Lord is in all His works of Creation!

God's wisdom is revealed in His arrangement of sections and segments, as well as in the number of grains. Each watermelon has an even number of stripes on the rind. Each orange has an even number of segments. Each ear of corn has an even number of rows. Each stalk of wheat has an even number of grains. Every bunch of bananas has, on its lowest row, an even number of bananas, and each row decreases by one, so that one row has an even number the next row an odd number. Such meticulous order did not derive from disorder.

The waves of the sea roll in on shore twenty-six to the minute in all kinds of weather. All grains are found in even numbers on the stalks. The Lord specified thirtyfold, sixtyfold and a hundredfold – all even numbers.

God has caused the flowers to blossom at certain specified times during the day, so that Linneus, the great botanist, once said that if he had a conservatory containing the right kind of soil, moisture and temperature, he could tell the time of day or night by the flowers that were open and those that were closed!

Thus the Lord in His wonderful grace can arrange the life that is entrusted to His care in such a way that it will carry out His purposes and plans, and will be fragrant with His presence. Only the God-planned life is truly successful. Only the life given over to the care of the Lord is safe. (Author unknown)

Resurrection of the Lord Jesus Christ

I love a good tract that lovingly and clearly presents the Lord Jesus Christ and the way of salvation. I have written a few. However, God wrote His own tract when He wrote the Gospel of John. He said in John 20:31, "But these are written, that ye might believe that Jesus is the Christ, the Son of God; and that believing ye might have life through His name."

Yet God seems to be declaring in Romans (1:1-4) that the ultimate proof that Jesus is God, is by His resurrection from the dead.

"Paul, a servant of Jesus Christ, called to be an apostle, separated unto the gospel of God (which He had promised afore by His prophets in the Holy Scriptures), concerning His Son Jesus Christ our Lord, which was made of the seed of David according to the flesh; And declared to be the Son of God with power, according to the spirit of holiness, by the resurrection from the dead."

Why is this? Because after declaring that Jesus would die for our sins, as millions of oxen and lambs portrayed when their blood was shed, God connected Jesus with these sacrifices.

John 1:29 declared through John the Baptist, as Jesus approached, "...Behold the Lamb of God that taketh away the sin of the world." Explicitly, the shed blood of these sacrifices pointed to Jesus, and Psalm 22, Isaiah 53, and many other passages speak of His death for us, as God spoke through the prophets in the Old Testament, hundreds of years before Christ came. Simultaneously, there were prophecies that said He would reign forever. Obviously, you can't have *both* without the resurrection.

Isaiah 9:6, speaking of Jesus, tells us, "For unto us a child is born, unto us a Son is given; and the government shall be upon His shoulder; and His name shall be called Wonderful, Counselor, The Mighty God, the Everlasting Father, The Prince of Peace."

In Jeremiah 23:5-6, God declares through His prophet Jeremiah, "Behold, the days come, saith the Lord, that I will raise unto David a righteous Branch, and a King shall reign and prosper, and shall execute justice and righteousness in the earth. In His days Judah shall be saved, and Israel shall dwell safely; and this is His name whereby He shall be called, THE LORD OUR RIGHTEOUSNESS."

(In 1 Cor. 1:3, Christ is our Righteousness.)

In Revelation 11:15, we are told, "And the seventh angel sounded; and there were great voices in heaven, saying, the kingdoms of this world are become the kingdoms of our Lord, and of His Christ, and He shall reign forever and ever."

No wonder Paul cried out, in 1 Cor. 15:19, "If in this life only we have hope in Christ, we are of all men most miserable."

Surely, the prior verse explains why, v17, "And if Christ be not raised, your faith is vain; ye are yet in your sins."

If Christ is not risen, the Prophets of the Old Testament were false. If Christ is not risen, the bloody sacrifice on Calvary was for nothing. If Christ be not risen, the New Testament apostles and disciples were deceived. If Christ be not risen, the Bible is untrue. If Christ be not risen, then He *lied,* because He told His disciples over and over again that He would rise from the grave. For instance, one of those times when He taught the disciples about His death, burial and resurrection, is recorded in Mt. 16:21. "From that time forth began Jesus to show unto His disciples, how He must go unto Jerusalem, and suffer many things from the elders and chief priests and scribes, and be killed, and be raised again the third day."

In fact, hundreds of years before in the Old Testament, God had stated in Psalm 16:10 (speaking of Jesus), "For thou wilt not leave my soul in Sheol, neither wilt thou permit thine Holy One to see corruption." This was reiterated and applied to Christ in Acts 2:27, after the resurrection of Christ. God records this prophecy and its meaning and fulfillment, more fully in Acts 13:35-37.

In this titanic battle between Christ and Satan for the souls of men and final and forever victory, even Creation itself has been impacted. After his fall, Adam was told by God in Genesis 3:17b, "…cursed is the ground for thy sake; in sorrow shalt thou eat of it all the days of thy life; v18, Thorns also and thistles shall it bring forth to thee; and thou shalt eat the herb of the field; v19, In the sweat of thy face shalt thou eat bread, till thou return unto the ground; for out of it wast thou taken—for dust thou art and unto dust shalt thou return."

Romans 5:12 highlights this cataclysmic event, "Wherefore, as by one man sin entered into the world, and death by sin; and so death passed upon all men, for that all have sinned."

Romans 8:21-22 startles us with this phenomenal announcement, "Because the creation itself also shall be delivered from the bondage of corruption into the glorious liberty of the children of God. For we know that the whole creation groaneth, and travaileth in pain together until now."

Yes, the whole world was at stake as the Lord Jesus Christ conquered death, sin, Hell, Satan and the very demons of Hell, in His glorious resurrection.

What a bloody mess the Lord Jesus Christ had been on the cross and even before, with the brutal treatment He had received. He had been blindfolded and smashed by callous soldiers. He was mocked, spit upon, His beard pulled out, taunted, beaten until it was quite possible His bowels were showing. The bowels of many did and they often died in the pre-crucifixion ordeal of being whipped into a mangled mass of torn and lacerated flesh and profuse loss of blood. He was crowned with a very painful wreath of sharp thorns.

God summed it up prophetically through Isaiah 52:14 hundreds of years before. "As many were astonished at thee; His visage was so marred more than any man and His form more than the sons of men."

It breaks my heart to think of Jesus suffering so much for me, for my sins, because He loved me. So it must have been also with Joseph of Arimathaea.

John 19:38 declares, "And after this, Joseph of Arimathaea, being a disciple of Jesus, but secretly for fear of the Jews, besought Pilate that he might take away the body of Jesus; and Pilate gave him leave. He came therefore and took the body of Jesus." Mark says he came "boldly" to Pilate. Matthew 27:57-60 tells us that Joseph, who was a rich man, buried Jesus in His own tomb, thus fulfilling prophecy. John 19:39 informs us that Joseph had help, none other than the surreptitious Nicodemus of John chapter 3, who had come to Jesus by night.

Jesus had told this religious but lost ruler that he must be born-again, which regeneration only occurs when Jesus is received as Lord and Savior. It *must* have happened. Read John 19:39-40, "And there came also Nicodemus, who at the first came to Jesus by night, and brought a mixture of myrrh and aloes, about a hundred pound weight. Then took they the body of Jesus and wound it in linen clothes, with the spices, as the manner of the Jews is to bury." (The passage goes on to say that it was a tomb in which man had never been laid.)

Apparently, both Joseph and Nicodemus had just come from the cross. Undoubtedly, their hearts were torn asunder as they saw the bloody, battered Jesus suffer and die for them, in their place. No more secrecy. No more holding back. No more caring about prestige, or position, or possessions. They committed themselves unreservedly to the Lord Jesus Christ, as anyone who really looks intently at Jesus on the cross will do.

Many of the disciples, however, were puzzled, devastated, confused. Think of the agony of Peter, who denied the Lord three times, as predicted, and went out into the night and sobbed bitterly and inconsolably when Jesus turned and looked at him. There was the interminable wait until the third day, which many of them, although told again and again by Jesus Christ Himself, did not yet understand.

The two disciples on the road to Emmaus furnish a perfect example of this. The episode is recorded in Luke 24:13-35. They were walking the 7½ miles from Jerusalem to Emmaus, talking about all that had happened, remembering that the third day had particular significance since His death. The risen Jesus came to them and walked along with them. Then Jesus, in Luke 24:27, continued His discussion with them, "And beginning at Moses and all the prophets, He expounded unto them in all the Scriptures the things concerning Himself."

When Jesus prayed and broke bread with them, they recognized Him and declared in v32-33, "And they said one to another 'Did not our heart burn within us, while He talked with us by the way, and while He opened to us the Scriptures?' And they returned to Jerusalem and found the eleven gathered together, and them that were with them."

Were they excited? Enough to walk another 7½ miles back to Jerusalem, for a total of 15 miles!

Nothing else matters when you have met the risen Christ! While they were telling their thrilling story, v36 says, "And as they thus spake, Jesus Himself stood in the midst of them, and said unto them, Peace be unto you." And in Luke 24:39, by way of assurance and confirmation, He declared, "Behold my hands and my feet, that it is I myself; handle me and see, for a spirit hath not flesh and bones, as ye see me have."

Here is a brief Biblical account of the resurrection of the Lord Jesus Christ. (Luke 24:1-7) "Now upon the first day of the week, very early in the morning, they came to the sepulchre, bringing the spices which they had prepared, and certain others with them, and they found the stone rolled away from the sepulchre. And they entered in, and found not the body of the Lord Jesus. And it came to pass, as they were much perplexed thereabout, behold, two men stood by them in shining garments; And as they were afraid, and bowed down their faces to the earth, they said unto them, why seek ye the living among the dead? He is not here, but is risen; remember how He spake unto you when He was yet in Galilee, saying, 'The Son of Man must be delivered into the hands of sinful men, and be crucified, and the third day rise again.'"

Of course, the Gospels of Matthew, Mark and John also carry the resurrection account. They were eyewitnesses. The Gospel writers must have been impressed by the huge 1- to 2-ton stone that had sealed the tomb. They all mentioned it. Josh McDowell adds these interesting observations, on pg. 233-34 of his book, "A Ready Defense."

Matthew 27 mentions that a large stone was rolled against the entrance of the tomb. The Greek Word for roll used here is "kulio." Mark uses the same word kulio, but in Mark 16 adds the preposition "ana," which explains the position of the stone after the resurrection. Essentially, this enlarged the meaning to "up, or upwards." It can mean, "To roll something up a slope or incline," which would mean that a slope or incline came down to the front of the tomb. Amazingly enough, Luke used the word "kulio," as well, but added a different preposition. In this case "Apokulio," means the stone was rolled not only away from the entrance, but since the women in this story used the word for the entire massive sepulchre, in context it points to a tremendous moving of this ponderous stone, carried away quite a distance, for free access to all.

Josh McDowell gives us a handy list of the appearances of Jesus after His resurrection, in his and co-author Bob Hostetler's book, "Beyond Belief to Convictions." See pg. 272.

1. To Mary of Magdala (Mk.16:9; John 20:11-18);
2. To the women returning from the tomb (Mt. 28:9-10);
3. To Peter (Luke 24:34; 1 Cor. 15:5);
4. To two followers on the road to Emmaus (Luke 24:13-33);
5. To the disciples and others (Luke 24:36-43);
6. To the disciples, *with* Thomas (John 20:26-29);
7. To seven disciples by the Sea of Galilee (John 21:1-23);
8. To five hundred plus followers (1 Cor. 15:6);
9. To James, His brother (1 Cor. 15:7); and to the
10. Eleven disciples at His Ascension. (Acts 1:4-9)

Dear friends, the founders of this world's religions are all dead—Buddha, Confucius, Muhammad, etc. Their bodies have long since decayed. Only Jesus Christ arose from the grave. Only Jesus had power over death. His tomb is *empty!*

When I was holding evangelistic meetings down in Big Bear country in California, I asked a Pastor/Lawyer friend I had met, Dr. Bob Topartzer, what would be conclusive evidence in any court of law that Jesus rose from the dead. The quintessence of what he said was that it had to be certain that Jesus actually died; it had to be true that He had been buried; and it had to be certain that He had risen again from the dead. He was *positive* that Jesus Christ had met all these tests, almost exponentially. So am I.

1. Jesus was dead. The centurion, an expert in those who were dead or just feigning death, checked Him out. They crucified hundreds. None lived. Jesus was dead.

He was taken down from the cross. Why? He was dead. Joseph of Arimathaea and Nicodemas handled His body, and cocooned Him in spices and linen, something they would never have done if He were not dead. The soldiers knew Jesus was dead, especially the one who had thrust a spear into His side and His heart; but all of them knew. He was dead. The disciples knew that He was dead. Mary, the mother of Jesus knew He was dead. While there was a flicker of hope, she would never have left her suffering son. The women knew Jesus was dead and later they brought spices to embalm His body. Jesus was dead.

The Jewish official certainly knew Jesus was dead, as did Pilate. They demanded and arranged a seal for His tomb. Jesus was dead.

2. Jesus was buried. Joseph of Arimathaea and Nicodemus knew that Jesus was buried. They tended to it, and it was Joseph's own tomb. Those ubiquitous

women knew He was buried. Mark 15:47 notices they "beheld where He was laid." Certainly, those brawny soldiers whose very lives depended on Jesus being buried and the tomb properly sealed and guarded, must have checked the body carefully and made very sure His body was in the Tomb. Their *lives* depended on it. They *knew* Jesus was buried.

3. Jesus *arose* from the Tomb. There is no other viable alternative. He was seen by His disciples, by some skeptics, including His own unbelieving brothers, who were convinced and converted. Doubting Thomas was thoroughly convinced when he met the risen Christ for himself, and bitter persecutor and enemy Saul was converted, transformed and broken by the love of Christ when Saul met Him on the Damascus road. Christ appeared over a period of 40 days, to individuals, to small groups and to over 500 at once. He was handled, touched, and ate with His disciples.

Jesus has met the test that would hold up in any court of Law. He was dead, He was buried and He rose again!

I wrote about the resurrection of Jesus Christ in my "Evidence" book, which God is using in hundreds of thousands of copies in Russian, Chinese, Korean, English, Romanian, and on the Spanish internet. At the risk of redundancy, I want to examine the main arguments skeptics, unbelieving seminary professors, atheists and enemies of Christianity have given against the resurrection.

The first objection states the disciples stole the body of Jesus. This was initially promoted by the chief priests and elders. The discombobulated soldiers, stunned and frightened, feared for their very lives, as the body they were supposed to be guarding was gone! The chief priests and elders gave them a way out, and even bribed them with money to *say* that the *disciples stole* the body while they slept. (Mt. 28:11-15)

Really now, if they were *asleep,* they could not have known that the disciples stole the body, a fatal contradiction in their own story. And *obviously*, it would have been incredible for all of them as professional soldiers, to have slept *simultaneously,* especially in view of the horrible death that awaited each if they failed. At this point, the disciples were shattered, scattered and afraid. Such a venture, impossible anyway, would simply have been beyond them. And *why* would they steal the dead body of Jesus Christ and *then* go out rejoicing to die for a lie?

The second objection is that the *soldiers* stole the body! *Why?* Because they wanted to die a horrible death (probably also by crucifixion), for stealing the body they were commissioned to guard? If they had stolen the body, these often persecutors of Christianity themselves, could have demolished the fledgling Christian movement simply by producing the body at any time.

The third objection is the "Swoon Theory." Jesus never did *really* die. He was placed in the tomb, revived, pushed the stone away and came out. We have dealt with this and shown that this is impossible. Jesus was bleeding profusely and unattended, with multiple injuries and a spear thrust into His heart. He was wrapped in linen and spices, which could have suffocated Him, and certainly would have imprisoned Him in a cocoon of hardening "glue," so that escape for a *powerful uninjured* man would have been impossible. He then had to move a mastodonic size stone with pierced hands, walk out on bloody mangled feet, past guards who would have instantly killed Him, and later convince His disciples that—He had conquered death. Absurd!

The fourth objection is that the disciples did not actually *see* Jesus; they only hallucinated, had a vision of Him and thought they saw Him. One unalterable fact—The Tomb was and is *empty*. The soldiers themselves reported the empty tomb, so they must have hallucinated too.

I remember dealing with a number of soldiers on drugs in Anchorage, Alaska. They described vividly to me what they saw. No two soldiers ever had the same hallucination at the same time. Thank God, I led many of them to Christ, and some of them became preachers of the Gospel.

However, Jesus appeared to more than 500 people at one time. The idea that they could all have the same hallucination at the same time about the same person is pure nonsense. Jesus appeared to people at nighttime, morning, evening—you name it. To believe in these cockamamie theories rather than the resurrection takes more faith than to believe the truth. Jesus Christ rose from the grave.

Another rather weak objection was that they went to the wrong tomb, a proposition proposed apparently seriously by a "theologian" named Kirsop Lake. Do I need to answer? God Himself forgot where the tomb was and sent His angels to the wrong place; Joseph forgot where his own tomb was, as did the soldiers, the women who saw Him buried, *and* the disciples?

One recent objection, many faceted in its approach, claims that the Gospel was stolen from the Gnostics, partly from the Nag Hammadi, and other sources. Some mythical "gods" of Greece and Rome, such as Zeus, Thor, had also supposedly been resurrected. I used to read the mythology of Greece and Rome for fun when I was in high school. I remember Thor was one of my heroes. "The God of Thunder rode to war, upon his favorite filly. I'm Thor, he cried, the horse replied, 'You forgot your thaddle, thilly!'"

No wonder!

Those who have read Greek mythology would have real trouble finding any correlation between the fancies of these fables, and the facts of Christianity.

In this same vein, Christ is supposed to have been married to Mary Magdalene, had children and apparently sort of *disappeared* into the ozone.

One of the most virulent of these attacks is "The DaVinci Code," written by Dan Brown. Among many other claims, is one that the tomb of Jesus has been found with His bones still in it, as the ossuary is labeled with that name and the name of "Miriam," and also several of the names of the disciples. I saw the replication and believe me, it takes a lot of imagination to decipher that group of names, not to mention the fact that scholars have shown that there were many living at the same time of the crucifixion named Jesus, as it was a popular name. Also common, were the names of some of the other disciples.

Most of these objections are done for sensationalistic purposes, and of course, to make money. Christian scholars have already decimated these claims, proving them false, exaggerated, distorted, and sometimes, outright lies. For one of these books, read "The DaVinci Deception," by Dr. Edwin Lutzer, and "The Case for the Real Jesus," by Lee Strobel. You will be glad you did.

Think with me for a moment. The crowd at Jerusalem during Passover was estimated as perhaps 2,000,000. The Tomb was open and empty. Anyone could go see it for themselves. In all of the bitter attacks against Christianity, *not one skeptic, not even the most truculent, ever dared say the tomb was not empty.* They, the crowd, the disciples, everyone, for hundreds of years, dared not!

They would have been depicted as fools. Only now, 2,000 years later, have we seen these ignorant, ill-advised attacks against the empty tomb, which is still in Jerusalem, and it is still *empty*.

It takes a lot of faith, and a lot of deliberate obfuscation, perhaps even willing ignorance, to believe these attacks.

One of the proofs of the resurrection is the sudden transformation of the disciples from mice to martyrs, dying and willing to die, often singing while they were burned, beaten and tortured to death for the resurrected Lord Jesus Christ. These same disciples had scattered simply at the arrest of Jesus, with Peter even denying he knew Christ. Something dramatic must have happened to cause them to transform into the men they became, fearless for Christ—to a man. That *something* was obviously the resurrection and their unswerving conviction that it had occurred.

Also, at the total risk of their lives, newly converted Jews who previously kept the Sabbath suddenly changed their worship day to Sunday, because this was *resurrection* day, and Christ was the first-fruits, as portrayed in Lev. 23:9-11 and explained in 1 Cor. 15:20.

When I first started writing, I discovered that some 66 million Christians had given their lives for the Lord Jesus Christ and now the total must be somewhere around 70 million. Even *today*, in Muslim countries and elsewhere, Christians are suffering and dying for Jesus Christ.

Absolutely nothing can explain the undeniable change in these disciples and converted Jews, but the bodily resurrection of the Lord Jesus Christ. No one ever cared for you, like Jesus. I am no Billy Graham, who has led millions to Christ, but I am thrilled that I have seen thousands trust Him, some in mass meetings and hundreds more in personal evangelism. Again, God did it.

Dear friend, I have given my life that people may come to know, live for, and glorify Jesus Christ. I cannot stand the thought of anyone going to Hell without tears welling up in my eyes, and however inadequately, I have pursued sinners for about a million miles, by car, boat, plane, train and even on foot. My wife Virginia has passionately sought souls along with me for over 50 years.

Let me share with you how you can be saved, sure of Heaven, and know it, 100%. God said so. 1 John 5:13, "These things have I written unto you that believe on the name of the Son of God; that ye may *know* that ye have eternal life, and that ye may believe on the name of the Son of God."

Romans 3:23 tells us "For all have sinned, and come short of the glory of God."

There are *no* exceptions. We are sinners because of the sin nature we are born with. We are sinners because we choose to sin. We are sinners because we go our own way and act as if we were our own God. We are sinners because of unbelief, which is really the root of all sin.

We are sinners because we have broken God's Holy Commandments. We have offended His holiness. Whether we have hated someone, which Jesus said was murder, looked on a woman or man to lust after them, which Jesus declared was adultery, whether we have lied (Jesus made it plain that liars do not go to Heaven), stolen, or broken any one of the Ten Commandments. (Ex. 20) God declared in James 2:10, that to break one of them is to break *all* of them.

We are guilty. If we have put anything or anyone before God, we are guilty. Isaiah 53:6 says tersely, "All we like sheep have gone astray; we have turned everyone to his own way; and the Lord hath laid on Him (Jesus) the iniquity of us all." We must realize we are *lost,* one breath, one heartbeat from Hell, before we can be saved.

Romans 6:23 adds "For the wages of sin is death; but the gift of God is eternal life through Jesus Christ our Lord."

Happily, Eph. 2:8-9 wafts this fragrance of love to our miserable condition, "For by grace are ye saved through faith; and that not of yourselves; it is the *gift* of God, *not of works,* lest any man should boast."

Once saved, Eph. 2:10 becomes our mantra!

John 3:3 informs us through religious but lost Nicodemus, as millions are today, that religion is not enough, good works are not enough, but as Jesus told

Nicodemus, "...Except a man be born again, he cannot see the kingdom of God."

John 1:12 tells us how. "But as many as received Him, to them gave He the power to become the children of God, even to them that believe on His name."

Our basic problem is that we are *not* children of God, or God would not ask us to *become* the children of God. We are the offspring of God, created by God, but spiritually we are not the children of God, until we receive the Lord Jesus Christ and are born-again into His forever family!

Romans 10:9-10 shows us what we have to know, to believe and receive to become His. "That if thou shalt confess with thy mouth the Lord Jesus (Jesus as Lord), and shalt believe in thine heart that God hath raised Him from the dead, thou shalt be saved."

"For with the heart man believeth unto righteousness; and with the mouth confession is made unto salvation."

Dear friend, are you now convinced that you are a lost sinner, without hope, except for Jesus? Do you believe that the Lord Jesus Christ, God manifest in the flesh, died for you, shed His blood for you on the cross? Do you believe with all your heart that He rose from the dead, loves you, and wants to save you? If so, you are ready for salvation.

Romans 10:13 gives us the key, which actually is just acting on your belief. "For whosoever shall call upon the name of the Lord shall be saved."

For two reasons we know this is incontrovertibly true. He loves you and would never die in bloody agony for you, and then turn you down, if you called with all your heart upon Him to save you. He is God, and He cannot lie, and if you call and trust on Him from your heart, your innermost being, He has committed Himself, He cannot lie, He *must*, and He will save you!

Please pray this prayer, or one like it. "Lord Jesus Christ, I am a sinner. I am lost. Please forgive me my sins by your shed blood, come into my heart and life and save me forever. I want to be born-again into your family and receive your gift of everlasting life. I will confess you, obey you and follow you, by your grace, the rest of my life. I repent of my sins and depend on you alone to save me, Lord Jesus."

Then, after looking at John 3:36, thank Him by faith for saving you and making you sure of Heaven.

You will then need to find a group of believers to associate with, love, worship with and grow in Christ with, along with obeying Him in baptism at your first opportunity, as He commanded. *Share* Him with others. (Luke 12:8-9)

Consider the choice, the consequences, and the cost, especially as highlighted in the next several chapters.

Committing Intellectual and Spiritual Suicide

L et me make it clear that all men are sinners by nature and choice, according to the Bible (Rom. 3:23) and certainly by human experience. It probably baffles the atheist who sometimes is living a better life than some professing Christians are; who works hard, lives a good, moral life and takes care of his family; only to discover that he is considered a sinner. This is especially true if he is not immoral, a sex addict, on drugs, a racist, hateful and he's compassionate to the poor, etc.

First, many *professing* Christians are not Christians at all, sad to say. I know, I was once like that, until Jesus really saved me, a professing but *lost* "Christian." God says of some of these, in 2 Tim. 3:5, that they are characterized by, "Having a form of godliness, but denying the power thereof; from such turn away."

In Titus 1:16, He is even more explicit, "They profess that they know God; but in works they deny Him, being abominable, and disobedient, and unto every good work reprobate."

Most damning of all to those who claim to be Christians but continue to go their own way, is 1 Cor. 6:9-11, "Know ye not that the unrighteous shall not inherit the kingdom of God? Be not deceived: neither fornicators, nor idolaters, nor adulterers, nor effeminate, or abusers of themselves with mankind, nor thieves, nor covetous, nor drunkards, nor revilers, nor extortioners, shall inherit the kingdom of God. And such *were* some of you, but ye are washed, but ye are sanctified, but ye are justified in the name of the Lord Jesus, and by the Spirit of our God."

Christians are new creatures in Jesus Christ, with the Living Christ dwelling in them to transform them. They have a *new nature,* but since they still have the old nature of sin to contend with, Christians can still sin, but they cannot *live* in sin. 2 Cor. 5:17 states, "Therefore if any man be in Christ, he is a new creature, old things are passed away; behold, all things are become new."

1 John 3:9 thunders, "Whosoever is born of God doth not commit (Gk: habitually practice, continually) sin; for His seed remaineth in him; and he cannot sin, because he is born of God."

The confusion is compounded when some atheists act like Christians and some professing Christians act like atheists, e.g., in not really believing and acting on the Word of God, however vociferous their profession.

Atheists, even those with Christian backgrounds, do not seem to be able to understand that multitudes who claim to be Christians are not. (Check Mt. 7:13-14) Therefore, they tar all *"Christians"* with the same brush.

In fact, they may sometimes point to the Crusades as an example of the rapacious, bloody, murderous history of "Christians," when this was not a *Christian* undertaking at all, though some among them may have been Christians, but a predominately Roman Catholic led response to an *even more* murderous conquest by Islam some time before.

One outspoken atheist wrote in the Hattiesburg/American newspaper recently, that atheists were apparently better people than Christians, because statistics show that comparatively few atheists are in jail, but a much higher percentage of Christians are in prison.

Perhaps many of those in jail repented and became true Christians, and perhaps many more had a "jailhouse conversion."

In the Mayflower Compact, it was clearly stated that spreading the Gospel of the Lord Jesus Christ, as well as freedom to worship without being under a mandated State church, was their purpose.

Patrick Henry endorsed and emphasized this, when he wrote in 1776, "It cannot be emphasized too strongly or too often that this great nation was founded not by religionists, but by Christians, not on religion, but on the Gospel of Jesus Christ. For that reason alone, people of other faiths have been afforded freedom of worship here."

This information and some following truths were taken from a booklet on the internet, "Forsaken

Roots," but I also have many more historical facts about the founding of our Country in books David Barton wrote.

Page 1 of "Forsaken Roots" asks, "Did you know that 52 of the 55 signers of the Declaration of Independence were orthodox, deeply committed Christians?"

Page 2 informs us, "It is the same congress that formed the American Bible Society. Immediately after creating the Declaration of Independence, the Continental Congress voted to purchase and import, 20,000 copies of Scripture for the people of this nation."

George Washington said in part, on Sept 19, 1796 (pg.7), "It is impossible to govern the world without God and the Bible."

John Adams, our second President, was chairman of the American Bible Society. On pg. 10 of "Forsaken Roots," he observantly and astutely comments, "Our constitution was made only for a moral and religious people. It is wholly inadequate to the government of any other." How true that has proven to be. The further we get away from God and the Bible, the more corrupt our government, society and schools become, so that now we have virtually no standard of right and wrong, and tremendous moral deterioration is taking place.

On pg. 14 of "Forsaken Roots," we find that in 1782, the United States Congress voted this resolution: "The Congress of the United States recommends and approves the Holy Bible for use in all schools."

The McGuffey Reader, a powerful Christian book, was used in our schools for over 100 years. Over 125 million copies of this book were sold.

Hospitals worldwide were largely a result of Christians and the Christian message.

Page 16 informs us that, "Of the first 108 universities founded in America, 106 were distinctly Christian, including the first." On pg. 19, "For over 100 years, more than 50% of all *Harvard* graduates were pastors." We owe our Constitution, our wonderful and free American Nation, and our educational and judicial system to the Bible, Christians and the Christian faith.

Yet a godless nation, Russia, inundated by Atheistic Communist, provides a stark contrast between atheism and Christianity. With the collapse of Communism and the chaos, corruption and criminal activity, which were unleashed, many Russians became Christians. When the rigid dictatorial regime glued together by fear, *disintegrated*, they turned to Christ.

While our schools are pandering to the Atheists and Agnostics, lest we offend some, and doing everything possible to get God out of the schools, according to Olga of the Kindness Foundation, Russian schools are now asking that *Christianity* to be taught in their schools.

Again, my atheistic, agnostic and infidel friends, what pure faith you must have to continue your belief that atheism is *better*—than *what?* You head blithely and nonchalantly for the abyss, truly, as the Bible says, in effect, "The god of this world has blinded your eyes, and your minds." America is the most wonderful country in the world's history! Millions have fought, and are fighting to get in and taste of the freedom we have here. America was founded and largely preserved by *Christians*. Show me an atheistic country with that kind of record.

(I realize that there is a difference between an atheist, who says there is no God and an agnostic, who says he does not know if there is a God or not. As well as an infidel who may believe in some kind of God, but does not believe in the Lord Jesus Christ, but I have lumped them pretty much together in writing this book).

It is only in the last 50 years or so, as the Bible has been dismissed from our schools, along with Christian principles and the Ten Commandments, that we have become a Nation under siege. This has spawned immorality, pornography, promotion of gay life styles, drugs, materialism, rebellion, lying, cheating, undisciplined school children terrorizing their teachers, abortion, hate, murder, alcohol abuse, brutality, rape, and hopelessness. Not to mention an epidemic of STD's, Aids, Pedophile paranoia and unfettered cruelty.

America and her schools are slowly but surely getting out of control.

Yet atheists cannot see that their insistence on removing God from our schools, and to some extent, our society, is the direct cause of this inchoate miasma. In fact, they want to go much farther and much faster on this path to destroy the very freedoms they now enjoy.

What faith…and folly!

This is intellectual and moral suicide, and when atheists have to face the God they say doesn't exist, it will be spiritual suicide, which will doom them forever.

The fact that they do not believe in God, does not affect His existence whatsoever. Just like a man who says he doesn't believe in gravity, and jumps out of a plane at 20,000 feet without a parachute. *Gravity* is not affected at all by his unbelief and he will find out soon enough that he was wrong.

Let us take a brief look now at Nature and Intelligent Design. Since I did not set out to write a lengthy book, I hesitate to do more than state some obvious facts. First, some atheists claim they do not see God in Nature. Astonishing!

An old but useful illustration concerns a man walking along and finding a watch, partly buried in the sand. He looks at it carefully, and with his analytical, scientific mind, expostulates, "Wonderful! What a marvel of evolution. Once when the earth was too close to the Sun, flames leaped out from the Sun. It was so hot, bits of metal seeped out of the mountains. As millions of years went by, wind drove the metal over the rough rocks down the mountain, and thus these little gears were joined together. Simultaneously, melted sand, slowly but irrevocably, formed the glass which eventually fitted over the gears. How nice that the two slivers of metal joined together to form the hands, that now enable us to tell time. Isn't evolution wonderful?"

Would that be the reaction of anyone in their right mind? Someone made the watch. There *had* to be a watchmaker. What would you think about someone who finds a watch, denies that anyone made it, and spends years trying to find out how it made itself?

Frankly, it takes a lot of faith to believe in Evolution. Imagine believing that a "Big Bang" explosion created our complex world, much like an explosion in a junkyard *creating* a 747. Explosions create chaos, not intelligent design. It is even more ridiculous to imagine that life came from non-life!

Or that chance mutations, almost always harmful, brought about the present evolution/creation of human life. Imagine believing that there is universal law, such as gravity without a lawgiver, or design without a designer, or a delicious, well-cooked meal without a cook. Truly, as quoted in the first Chapter, God has made foolish the wisdom of this world.

The answer is very simple. Gen. 1:1, "In the beginning God created the heaven and the earth."

Even in Creation, God has shown His marvelous love and care for us. As the late Dr. James Kennedy, former pastor of Coral Ridge Presbyterian Church in Fort Lauderdale, Florida, with a tantalizing alphabet of earned degrees adorning his credentials, said in his fine book "Why I Believe" pg. 42 "...without the moon, it would be impossible to live on this planet. God has provided the moon as a maid to clean up the oceans and shores of all our continents." He adds on pg. 43, "Then there is the amazing nitrogen cycle. Nitrogen is extremely inert. If it were not, we would all be poisoned by different forms of nitrous combinations. However, because of its inertness, it is impossible for us to get it to combine naturally with other things.

Yet,—it is definitely needed for plants in the ground. How does God provide to get the nitrogen out of the air into the soil? He does so by lightning! One hundred thousand lightning bolts strike this planet daily, creating a hundred-million tons of usable nitrogen plant food in the soil every year.

Forty miles up is a thin layer of ozone. If compressed it would be only a quarter of an inch thick, and yet without it life could not exist. Eight killer rays fall upon the planet continually from the sun, without ozone we would be burned, blinded and broiled in just a day or two. The ultraviolet rays come in two forms: longer rays, which are deadly and are screened out, and shorter rays that are necessary for life on earth and are admitted by the ozone layer.

Furthermore, the most deadly of these rays are allowed through the ozone layer in just a very thin amount, enough to kill the green algae, which otherwise would grow to fill all the lakes, rivers and oceans of the world. How little we realize what God is continuously doing to provide for our life."

This is not to mention His provision of oxygen, and especially water, without which we could not live. We are the only *known* planet in the universe that has a sufficient amount to sustain life. Yes, even nature says, "I am God, and I love you." But nothing manifests His love toward us so much as His only Son, Jesus Christ, torn, spit on, rejected, battered, beaten, shedding His blood for us, dying in our place, to pay for our sins, and purchase us from the horror of Hell.

Notice some of His handiwork. If the axis of the earth moved an appreciable number of degrees one direction, we would all freeze to death. If it lurched the other way, we would be burned to a crisp. The same thing is true regarding the exact distance we are from the Sun, not too close and not too far.

The Earth at the Equator rotates about *1,000 mph.* The Earth orbits the Sun at about 67,000 miles per hour—and you are on it. As Moody Adams says in his book "Proof" on pg. 27, "If God is dead, who is flying this plane?"

On the same page, Adams notes, "We are sitting on a spaceship called 'earth' in a galaxy that is flying through space at 1.1 *million* miles an hour. Inside this fast-flying galaxy, we are rotating around its nucleus at 500,000 miles an hour. Some folks seem to prefer to think that God is dead, or doesn't exist; therefore they are free to *do their own thing.*"

On pg. 29, Moody Adams makes a pithy observation. "It makes me think of what would happen on an airplane if the little flight attendant ran out of the cockpit shouting, 'The pilot is dead, and the co-pilot is dead. We can do our own thing. We are free.'" Oh?

In his book "Proof," Moody Adams records on pg. 35 that George Gallup said, "I could prove God statistically. Take the human body alone. What are the chances that all the functions of the human body would just happen?

The body grows from a single cell to 1,000 billion cells, each doing its fantastic job. If these cells were placed end-to-end they would circle the earth over 200 times. The skeletal framework is made up of 206 bones. These are more durable and longer lasting than steel. There are 39 skull bones, 26 spinal vertebrae, 24 ribs, 2 girdle bones and 120 other bones scattered throughout the body.

In an average man, they only weigh 29 pounds. The joints are self-lubricating. The skin covering your body has two million sweat glands and 10-15 thousand million centers, which report to the brain. These are a remarkable protection against heat, cold, sun, water, disease and injury."

Yet every one of us began from a *single* human cell. Incredibly, many of these cells are so small that a million of them could be put in a space no larger than a pinhead. With its inimitable DNA, each cell is a micro-universe, of almost incomprehensible complexity.

Jonathan Sarfati, in his book "Refuting Evolution," on pg. 122, records an illustration that Biochemist Michael Behe uses in his powerful anti-evolution book, "Darwin's Black Box."

He gives this mousetrap illustration to illustrate that on a practical level, information specifies the many parts needed to make machines work. Behe calls this "irreducible complexity." He gives the example of a very simple machine—a mousetrap. This would not work without a platform, holding bar, spring, hammer and catch, all in the right place. If you remove just one part, it won't work at all—you cannot reduce its complexity, without destroying its function entirely. The thrust of Behe's book is that many structures in living organisms show irreducible complexity, far in excess of a mousetrap or indeed any man-made machine.

For example, he shows that even the simplest form of vision in any living creature requires a dazzling array of chemicals in the right places, as well as a system to transmit and process the information. This, in itself, is the death knell to evolution, since no matter how much time is given (time is one of the unstated "gods" of evolution); the stubborn and unalterable mousetrap facts *remain* the same.

On pg. 123-124 of Jonathan Sarfati book "Refuting Evolution," he quotes Molecular biologist Michael Denton, who writes as a non-creationists skeptic of Darwinian evolution, and explains what is involved.

"Perhaps in no other area of modern biology is the challenge posed by the extreme complexity and ingenuity of biological adaptations more apparent than in the fascinating new molecular world of the cell. To grasp the reality of life as it has been revealed by molecular biology, we must magnify a cell a *thousand million* times until it is twenty kilometers in diameter and resembles a giant airship large enough to cover a great city like London or New York.

What we would then see would be an object of unparalleled complexity, and adaptive design. On the surface of the cell, we would see millions of openings, like the portholes of a vast space ship, opening and closing to allow a continual stream of materials to flow in and out. If we were to enter one of these openings, we would find ourselves in a world of supreme technology and bewildering complexity.

Is it really credible that random processes could have constructed a reality, the smallest element of which—a functional protein or gene—is complex beyond our own creative capacities, a reality which is the very antithesis of chance, which excels in every sense anything produced by the intelligence of man? Alongside the level of ingenuity and complexity exhibited by the molecular machinery of life, even our most advanced artifacts appear clumsy."

Some years ago, Dr. Henry Morris, a scientist and a Christian scholar, teamed with the inestimable theological intellect of Dr. John Whitcomb, to slow the tide of the then largely unopposed evolution tidal wave engulfing America, and wrote the humongous book, "The Genesis Flood."

This was a shot across the evolutionist bow. Dr. Morris wrote many other books, including one called "The Twilight of Evolution." Dr. John Whitcomb wrote on "The Early Earth." Ken Ham of "Answers In Genesis," wrote and spoke all over America. Dr. John Morris, the son of Dr. Henry Morris, and a scientist and scholar in his own right, now heads up the Institute for Creation Research, his father once led.

I have been in the homes, and/or eaten with, or spoken on the same platform with all of these men except Dr. Henry Morris, and I at least had the honor of meeting him and reading his books. Dr. Ken Ham has written "The Lie," speaking of Evolution. Dr. Duane Gish, of the Institution of Creation Research debated evolutionists everywhere. Dr. Josh McDowell, who wrote irrefutable books on apologetics for the Bible, spoke at, and debated, skeptics, college professors, or evolutionists, on over 1000 colleges and universities, worldwide.

Since the evolutionists more and more, don't dare face really knowledgeable Creationists, they are trying desperately to win politically by declaring that "Evolution is a proven fact, and that all Creation teaching should be banned from our educational system." Although in the majority, they are running scared. They shudder at the thought proposed that both evolution and Creation should be fairly taught, with the student allowed to make the choice—so much for freedom of speech!

Years ago, in a church I had started, and was the Pastor of in Anchorage, I had a lovely 17-year-old ball of fire for Christ, a girl of Philippine descent named Hannah, whom I affectionately called "Hannah Banana." She was being bombarded by evolution in her large high school, and apparently, the school issued a challenge for anyone to come and present the Creation/Biblical point of view. Hannah asked me to take the challenge. I had just had a heart attack and was weak as a cat, but thought I could summon the energy for one 30-40 minute period.

To compound my problem, I forgot my heart medicine when I left for the school. I spoke, and the students, who had been prepared and coaxed by the teachers on additional questions, began their inquiries. Some of the teachers did as well.

To my utter consternation, after I finished the first class with its encounters, I was told that the appointment included speaking and answering questions, in every class, all day, with a grand finale in the auditorium for all students and teachers as the last gladiatorial contest. The arena was prepared and the lions were about to be let loose.

Reminds me of the quip that the games at the Coliseum had to be closed down because the lions were eating all the prophets…

I sent Steve, a young football player we had led to Christ, to my home for some heart medicine, and held on until he got back. He was a Samoan, a tiger for Christ, who later played some football I believe, at the Air Force Academy, until injured.

I called with all my heart on the Lord to empower me and speak through me to His glory. I was totally unprepared for the day long battle, but I trusted God to work to His glory, even if I made a fool of myself. I was determined to be strong, no matter how ignominious the experience might be. Talk about Custer's last stand.

In all of the classes, God enabled me to meet the questions fairly, but much more than that, to speak of the revealing of God in Creation and in His Son, Jesus Christ.

Then it was time for the auditorium event. Almost none of the students seemed to be on my side, or they were too timid to speak up. The same was true of the teachers; they touted the "proofs" of evolution, and asked loud and clear if I believed Jesus Christ was the only way to Heaven. I said "Yes" unequivocally. He said so in John 14:6, "…I am the way, the truth, and the life; no man cometh unto the Father, but by me."

The auditorium erupted in an uproar. Some yelled, "What about the Hindu people, what about the Muslims, etc."

I countered, "Why would God sacrifice His own Son in bloody agony, if any of these other gods would do?"

(It is *rejecting* Jesus that will deny access to Heaven. Moreover, God is on record declaring that He is not willing that *any* should perish. (2 Peter 3:9b) That is why thousands of missionaries have given their life's work, and often their life's blood, to spread the Gospel to all without Christ, whatever their *religion* might be. One of my grandsons is currently in a Muslim country reaching dear lost people for Jesus Christ.)

Jesus did not come to the earth He created, to be just another god on the *god-shelves* of humanity. India alone has millions of gods and there are many false gods in other nations. Jesus was the one and only true God made manifest in human flesh. Besides, how can one claim to be a Christian, and deny what Jesus said? What infernal juxtaposition to call Him a liar and your Savior. Impossible! No liar can get you saved or get you to Heaven.

Jesus came not only to save men from their sins, but from the false impotent "gods" that "plague the world."

Remarkably, all other "gods" are imaginary.

Occasionally, some human claims, or is believed by some to be God, but those who are now dead are rotting in their tombs. Only Jesus conquered death and rose from the tomb. There is no point in following a loser. The Tomb is still *empty*. Jesus is still God.

Finally, on this subject, when the uproar subsided a little, I reminded them that the claim was not one I made. The Lord Jesus Christ Himself made the claim. As a true Christian, I have no choice but to believe it. So the argument was not between me and them, but between them and Jesus.

I was aware of a teacher from one of the classes that had asked some penetrating questions and seemed friendly, but the Holy Spirit seemed to warn me inwardly. The teacher was intelligent, but seemed unctuous in his demeanor and speech toward me.

Sure enough! Part of the program seemed to be to convince the students that evolution was a fact, and Creation was a religious myth. It was not an *unbiased* search for the truth.

In the midst of that great ocean of students and teachers, this teacher stood up and declared, "Mr. McElveen, evolution is a fact! The Nobel Prize was recently awarded to 72 evolutionary scientists who declared this fact and *proved* it."

I said, "Mr...., do you remember the Piltdown Man, who was lauded as a real missing link according to evolutionary scientists? This find was publicized about 1912, and continued being taught as an evolutionary marvel into the early 1950's, as a prime example of evolution. Some of us studied about him in college. Some 500 scientists wrote their pontifical doctoral dissertations on this evolutionary wonder.

It turned out to be a *hoax*. It was the jaw of an ape and a human skull artificially aged. If 500 scientists can be wrong, so can 72!"

The silence was deafening. The teacher sat down. Realistically, evolution comes much nearer being a farce than a fact.

After the day, I was told that at least one student had accepted Christ as their Savior, and that it had made an impact on the school, particularly in getting Christian students to stand up for Christ unashamedly.

I had simply decided if Christ could be a fool, hanging naked and blood drenched on a rugged cross, for me, I was willing to be a fool for Him. I love Him.

I tell this true story to encourage others. He will work through you if you step out by faith.

The Challenge—Trust in Bones or Bible?

God's love gift to you is Jesus and His salvation.

Even though God loves you, and sent His Son Jesus to the cross to die in excruciating agony for you as He shed His blood for your forgiveness and salvation, we need to mention a fact seldom mentioned today. Look carefully at John 3:36, "He that believeth on the Son hath everlasting life; and he that believeth not the Son shall not see life; but the wrath of God abideth on him."

Every moment, day and night, outside of Christ, living for yourself, breaking God's law, you are one breath, one heartbeat, from Hell. Forever!

In sharing with atheists, I may not mention Bible evidence at first, but simply give the Gospel to them with prayer and a broken heart, and I have seen some converted this way. After all, it is the Holy Spirit acting on the Word of God, which does the work. Facts can simply tear down the barriers they are hiding behind and usually evolution is one strong barrier.

I remember a young agnostic, brought to me by a wonderful young lady named Carmen. He was adamant and proud of his either agnostic or atheistic position, and though Carmen was a terrific witness, she couldn't get anywhere with him. We sat down at a table in Anchorage, Alaska and although answering a few questions, primarily I just talked to him about the wonderful Lord Jesus Christ, His death, burial and Resurrection. Then I asked him if he would like to receive Christ as his very own personal Lord and Savior. He said "Yes," and began to pray. Suddenly he jumped straight up and cried, "I'm born-again, Bro. Mac, I'm born again!" And he was, later spending some time overseas as a missionary.

On the other hand, if an atheist is hostile and unresponsive, or aggressive and obnoxious, I pray for him and then ask if I may have two or three Bible studies with him. Depending on the obstacles he raises, I center on proving that the Bible is the Word of God, proven by prophecy, etc., that Jesus Christ is God, that He died on the cross, shedding His blood for us, and rose again from the dead. I may deal some with evolution if that is a problem.

I remember a couple in Anchorage, Alaska. The man was some kind of Heavyweight champion in the Navy, or had been, and was married to a beautiful, but atheistic wife. He claimed to be an agnostic. How God worked!

What a joy when they both received Christ and began rejoicing in His sweet salvation. I remember getting a joyful call from them some time later from Rancho Cucamonga, California.

Sir Fred Hoyle was an eminent, award-winning British scientist, who has taught as an internationally recognized astronomer and mathematician in universities in England and America. Chandra Wickramasinghe is another British scientist who is an authority on interstellar matter. He headed the mathematics-astronomy department at University College in Cardiff, Wales. As reported in the Times-Advocate in Escondido, California, they have declared the "Darwin Evolution Theory, Absurd."

They array the findings of microbiology, mathematics, computer technology and the fossil record against the Darwin theory. They declare that the chances of random chemical shufflings in some primordial soup producing the complex basic enzymes of life are only one to 10 to the 40,000 power, or one followed by *40,000 zeros!*

These scientists make no claims of being Christians, but acknowledge that the bio-molecules necessary for life are so exceedingly complex that outside intelligence for explicit instructions were required. They claim that the "scientific world has been bamboozled into believing that evolution has been proved," adding that, "Nothing could be further from the truth." Speaking of scientists in general, Sir Fred Hoyle suggests that they dare not denounce the Darwin evolution theory as it would endanger their careers, and shatter everything they have been taught. It might even open the door to ridicule and rejection among their colleagues. He believes that many of them *know* that Darwin's theory of evolution is not true.

How tragic that we are becoming more and more forced to let our children be taught this fallacy; then see more and more of them turn away from Christ and the Bible. It is bad enough in high school, but colleges and universities are allowed to leave the Christian message in shambles with their Geological Column, and assumptions.

They snow the students, albeit some of them perhaps innocently as they were thus indoctrinated themselves, with the Era divisions, Paleozoic, Mesozoic and Cenozoic. Not to mention their evolutionary armamentarium of Cambrian, Ordovician, Silurian, Devonian, Carboniferous, Permian, Triassic, Jurassic, Cretaceous and Tertiary time periods, supposedly set in rock, pardon the pun.

To some extent, evolutionists judge the age of the geological strata by the fossilized animals it contains, and sometimes judge the age of the fossil deposits by the geological strata in which it is located. Circular reasoning at its worst!

In a conversation with Dr. John Morris, who has written revealingly on the Mt. Saint Helens volcanic eruption, he told me that after he and his team examined the site, they discovered that layers were deposited in *hours* that scientists had been saying it took *millions of years* to form.

Another anomaly that puzzles evolutionist to this day is why there are fossils of sea creatures on the mountaintops of the world, particularly Tibet. The only answer is the Biblical answer of Noah's flood, and this flood and its catastrophic results are carefully delineated in "The Genesis Flood" by Morris and Whitcomb.

While all men are sinners, as declared by God in Rom. 3:23, and have broken God's holy law, the Ten Commandments, and offended His holiness irretrievably—short of repentance and faith in the Lord Jesus Christ—atheists have gone a step farther. Many sinners have not knowingly denied God, or Jesus Christ, or emasculated the Bible deliberately. When shown the Word and given the Gospel, many of them gladly repent and accept Christ.

Atheists have taken the extra step and denied that there is a God, denied that the irrefutable Bible is true and ignored the mountains of evidence it contains. They consider Jesus Christ to be only a *human* teacher, if He existed at all, and have scoffed at the cross and the incredible suffering that Jesus Christ did for them.

God help them. Pray for them and love them. It may be the last love they ever receive, here or hereafter, when a merciful God has had enough!

In Alaska, after starting a church in Cooper Landing and holding evangelistic meetings everywhere, my family and I moved to Anchorage to start a church there.

I invited an atheistic Jew into my home, who headed up the TV Station in Anchorage, as he had previously done in Hawaii. He gave stinging attacks on

Christianity, and was very good at what he did. He was not just an ordinary genius, but a very high ranking member of Mensa, at the top of the genius list. He was widely known as the "Hairy Hornet."

Surprisingly, we had a very good time together as he visited with my wife, Virginia, and I. After several sessions, I had to move on to other fields, but later discovered he had *joined* the Presbyterian Church.

I made it clear to the Hairy Hornet, that if he was right, I had lost nothing, but if I was right and he was wrong, he would spend eternity in Hell.

Virginia and I gave him the way of salvation clearly, and then I challenged him to honestly take this test. I even made out a contract and he signed it. This is the gist of the contract and prayer.

I said, "If you are serious about knowing if God really exists, and whether Jesus Christ is really God manifest in the flesh, please read this verse and take this challenge." John 20:31, "But these things are written, that ye might believe that Jesus is the Christ, the Son of God and that believing ye might have life through His name." That is *why* the book of John was written.

The Challenge!

Read one chapter of John every day until you finish the book of 21 chapters. Pray this or a very similar prayer every time.

"God, if there really is a God, and if Jesus Christ is Messiah, God manifest in the flesh, convince me deeply, and I will accept Jesus as my Lord and Savior. I will love Him, follow Him and obey Him all the days of my life. I will ask the living Lord Jesus Christ to cleanse me by His shed blood from all sin, as I now repent and turn from them to Him. I will ask to be born-again by receiving Him and becoming a child of God. I will receive Him by faith, trusting in Him alone for my salvation. I will then thank Him for His forever salvation, and know that I now have everlasting life with Him in Heaven forever." (John 1:12, 3:3, 3:36) If you aren't sure you are saved, I urge you take this challenge.

SIGN HERE: _____

Dear atheist, skeptic, evolutionist friend, your attempt to ban creationists from debating you in public schools, or presenting their facts along with yours, squelches free speech. The only way that you can win against knowledgeable creationists, is by political pandering, becoming political sycophants, wooing liberal judges, or overwhelming the liberal media and the sympathetic educational system. But as Christians begin to stand up, and Creation Scientists take you on, your free ride may be over. We love you, and pray God will open your eyes, for your sake, your children's sake, for others you may destroy, and ultimately, for the lover of your soul, *Jesus!*

God Loves You and We Love You!

W e spoke of some of the Lord Jesus Christ's terrible suffering on the cross for you, for us. Yet we did not even touch on the most horrendous suffering this Old World has ever seen, for you. Beyond the physical suffering, the incomprehensible agony of Jesus, as God the Son was being separated from God the Father, as He who knew no sin, became sin for us on the cross. This is highlighted by His agonizing cry on the cross, as He bore our sins in His body on the tree, "My God, My God, Why hast thou forsaken me!" Creation was shocked, Darkness enclosed the horrible scene and the Sun hid its face, as Jesus paid the full price for our sins. Please don't turn down such love.

My beautiful wife of over 60 years now, told me a few months ago, that God had laid a tract on her heart. She wrote it up and our daughter Ginger dressed it up and we lightly edited it, and put in a picture; but it was her idea and her writing. I pray that all readers will enjoy it and that some will come to Jesus through it. Lives have already been touched.

Where Does This Train Go?

Big Mac Publishers

"Mr., Where Does This Train Go?"

By H. Virginia McElveen

D ad Ryan was in the hospital. He had a dream that was so real to him he could not get it off his mind. In the dream, he was about to get on a train. When he asked about the destination of the train, nobody knew, not even the conductor, or the ticket agent. They in turn asked him where he was going, and he answered, "I just do not know where I am going." Again, he asked, "Where does this train go?", but he got no answer. However, he had to get on that train!

When he woke up, he was very frightened. You see, he was a very good businessman, very precise in all the details. He was careful to lay out all the procedures in business planning. Nothing was left to guesswork. He had a son that had learned the business, and the son was just like his father, very meticulous and hard working. They were successful and the business had grown into a very large company. While Dad Ryan was still thinking about the dream, his son John came in to see him on his way to work. The hospital was close by the business, and he loved and was concerned about his dad, so he stopped by. Dad Ryan said, "John, I had a dream and I can't get it off my mind."

He told the dream to his son, and John said, "Dad, don't let it bother you. It was just a dream. Maybe it was because you are so sick."

Dad Ryan said, with deep feeling, "No, Son, I have to know the answer. Go get a preacher to tell me where I am going when I leave this old earth. That may be the message of the train in this dream."

"But Father..." objected John, but before he could continue, Dad Ryan countered, "Please Son, I have to know. Hurry and have him come tell me."

John, seeing the pathos in his father's eyes and hearing the urgency in his voice, said, "I will, Father."

John found a Pastor in a large, beautiful church and pressed his way past others to talk to him. When they got into the ornate office filled with a multitude of religious books, John told the Pastor about the dream and what his father wanted to know. The Pastor seemed somewhat taken aback. He said, "Young man, I do not know. I never tell my people how to die, or where they are going. I do tell them how to live. I tell them to be good, live right and they will be O.K."

John listened for a few minutes, with growing dismay. Suddenly, he jumped up and ran out of the office. As fast as he could, he located another Pastor and asked him, "Where would my father go if he should die?" This Pastor answered, "I know your father and you don't have to worry about anything. He has given a lot to this Church. Please don't be so alarmed! Dad Ryan is a good man."

John just couldn't take it anymore, but he had to find someone who could explain the dream to his father, so he went to yet another Church. Before he went inside, he sat down and tried to think through the dream. While he was sitting there, a young man named Bob walked up to him and with a friendly smile, said, "Good morning." John looked at him, but said nothing. Bob said, "May I ask you a question?"

"Yes," John answered, "You may ask me a question." Bob asked, "If you should die today, where would you spend eternity?"

John couldn't believe his ears. This could not be just a coincidence. He jumped up and pled with Bob, "Please, I beg you, get in my car and go to the hospital and tell my father and me at the same time, the answer to your question. He is so sick and needs to know where he is going before he dies!"

Bob got in John's car and prayed all the way to the hospital. When they arrived at the bedside of Dad Ryan, John declared, "This man knows the meaning of your dream, Dad Ryan." Bob said, "Let us pray together for God's leading and for the Bible's answer to your need, Jesus Christ."

Then Bob said, "Dreams won't lead us to Christ, but God has had mercy on you and alerted you to your need. The train is like life and all of us are on it. The train cannot tell you where you are going, but our life train has two eternal destinations. One is Heaven; the other is Hell, the Lake of Fire. All passengers will go to one of these places when they die. The reason you are not sure of your destination, Dad Ryan, is because you have never realized that you are a LOST sinner. God says so in the Bible. Romans 3:23 tells us, 'All have sinned and come short of the glory of God.' Romans 6:23 adds, 'The wages of sin is death, but the *gift* of God is eternal life through Jesus Christ our Lord.' All of us have broken the Ten Commandments of a Holy God. We have gone our own way, and done our own thing as if we were God. We all deserve Hell forever. But there is good news!"

Bob paused, and the eyes of Dad Ryan and John were riveted on him.

"Jesus made a way for us. In fact, He is the Way!" In Eph. 2:8-9, God says, "For by grace are you saved through faith, and that not of yourselves. It is the *gift* of God, *not of works,* lest any man should boast."

Bob continued, "Jesus Christ, God clothed in human flesh, the God-man, died on the cross in our place, and shed His blood on that terrible, painful cross, to pay for all our sins. No matter how 'good,' we think we have been, the Holy God sees us as lost, hopeless sinners, dangling each moment, and every breath over a lost eternity in Hell. Millions have been alive and vibrant one moment, and the next moment they have plunged into eternal torment with no hope forever."

"We sin because we have an old sin nature, all of us. We go our own way, a fatal mistake. Many are very religious, but lost. Many try to be good enough to get to Heaven, or try to help Jesus save them by doing religious rituals, or good works, which God says are only 'filthy rags'. Can you imagine hanging the filthy rags of our good works on the cross to 'help' Jesus save us? God wants to save us, and make us new creatures in Christ forever." (2 Cor. 5:17)

Then Bob addressed Dad Ryan personally. "Do you understand that you need Jesus to know where you are going?" The father, Dad Ryan, looked first at his son, John, and then at Bob. He said, "I believe that you have explained my dream. I need to know Jesus. What do I do?"

Bob explained, "God has arranged to forgive your sins through the shed blood of the Lord Jesus Christ. He has arranged to make you a born-again child of God, with a new nature. He says in John 3:3 that all of us have to be born-again to become children of God to go to Heaven. In John 1:12, He tells us how; by receiving Jesus, God's *gift* to us."

Bob looked at John and Dad Ryan, and they were looking very intently at him. "Actually," he continued, "you don't have to do anything. Jesus did it all on the cross. What He asks you and I to do, is simply to receive Him, call on Him and believe on Him, with all our heart."

Romans 10:9-10 says, "That if thou shalt confess with thy mouth the Lord Jesus, and shalt believe in thine heart that God hath raised Him from the dead, thou shalt be saved. For with the heart man believeth unto righteousness, and with the mouth confession is made unto salvation."

He continued, "Exactly what Jesus did to save us is concisely stated in 1 Cor. 15:1-4 and Romans 10:13, which shows us how to make saving application of these truths, transferring them from head-belief, to heart belief. 'For whosoever shall call upon the name of the Lord shall be saved.'"

Dad Ryan's voice was clear, but weak, "Bob, I now see my problem. I am a lost sinner, but I have known about Jesus for a long time. However, I never realized my deepest need. Bob, I want to pray, to call on Jesus with all my heart to save me. I do believe Jesus died on the cross for me and shed His blood for my sins, and I do believe He rose bodily from the tomb, and that He is God. And I know that I am a lost sinner." "Dad," John cried brokenly, "I see my need of Jesus and His salvation too. I will pray with you."

Two voices lifted as one to the Lord Jesus Christ, as Bob kept pace with "Amen, Praise the Lord," in a subdued and holy whisper.

Dad Ryan prayed, "Jesus, I ask you in your love and grace to save me and let me know my destination for sure. Wash me clean by your shed blood and make me a child of God."

His son, John was sobbing as he called on Jesus with his daddy, holding tightly to his daddy's hand.

Bob whispered, just after they called on Jesus, "God loves you so much, He would never turn you down if you believed with all your heart. He would not, cannot lie, if you asked Him to save you and believed that He did."

Both Bob and John heard his daddy, with great relief and joy in his voice, call out, "Thank you, Lord Jesus for saving me. I know my destination. I am going to Heaven to be with you forever."

John was so happy. He also thanked Jesus for saving him, and Bob pointed out 1 John 5:13 to him. As Bob quoted that wonderful verse, John felt the hand of his father slip slowly out of his hand. He had gone to be with his Savior.

"These things have I written unto you that believe on the name of the Son of God; that you may *know* that you have eternal life, and that you may believe on the name of the Son of God."

Bob and John wept with joy mingled with sadness that Dad Ryan was dead, but they rejoiced with unspeakable joy that now he knew and had arrived at his destination, Heaven with Jesus forever!

Dear Reader, I pray you will confess your lost condition and ask Jesus for forgiveness. Tell Jesus Christ you want to repent by calling on Him to save you and turn you from your sins. Ask Him to give you everlasting life. Trust Him with all your heart as your Lord and Savior, and He *will* save you. Please call on Him now.

Big Mac Publishers

Big Mac Publishers

"There Is No God"

Daily Reflection

2.1.09 By Mark D. Roberts, Laity Lodge Senior Director and Scholar-in-Residence. (http://www.thehighcalling.org/Library/ViewLibrary.asp?LibraryID=4949)

"For the wicked boast of the desires of their heart, those greedy for gain curse and renounce the Lord. In the pride of their countenance the wicked say, 'God will not seek it out.' all their thoughts are, 'There is no God.'" [Psalm 10:3-4]

Throughout Britain these days, buses are displaying signs paid for by donations to the British Humanist Association. The signs read: "There's probably no God. Now stop worrying and enjoy your life."

Not to be outdone, *America* had its own *anti-God* display recently. Last December in Olympia, Washington, an atheist group posted a sign next to a privately sponsored nativity scene. The sign read, "At this season of the Winter Solstice may reason prevail. There are no gods, no devils, no angels, no heaven or hell. There is only our natural world. Religion is but myth and superstition that hardens hearts and *enslaves* minds."

Though the forms of propaganda might be new, the enthusiasm of atheism is not. Ages ago, the Psalmist bemoaned the pride of the wicked, who boasted of their selfish desires and who were convinced that, "There is no God."

Therefore, when we see the boldness of today's atheists, we mustn't lose heart. *Rather, we must be confident in our faith,* clear in our communication, and conscientious in our actions. For, not only is the atheist creed false, *but it lacks the power to change lives.* Thus, we have the chance to persuade people of the truth of the Gospel, both through our words and especially through our lives. (End of Daily Reflection article)

*****

Atheists are putting up public signs that say that believing in God "enslaves" us. Jesus said in John 8:32, "And you shall know the truth and the truth shall set you free."

Sadly, the ones that are enslaved are the very same atheists who hang on by faith to an ill-supported theory simply because they cannot accept God. They have painted themselves into a position from which their pride may not ever allow them to escape. They have fashioned their faith and belief not on the actual evidence, but rather have twisted their selective evidence to fit their biased belief. That is not real science. It is blind faith of the worst kind.

Few have the "Faith" of an Atheist

Monday, February 09, 2009
(http://www.truthbook.com/news/labels/Rescuing%20Darwin.cfm

Half of Britons do not believe in evolution, survey finds. More than one-fifth prefers creationism or intelligent design, while many others are confused about Darwin's theory.

Half of British adults do not believe in evolution, with at least 22% preferring the theories of creationism or intelligent design to explain how the world came about, according to a survey.

The poll found that 25% of Britons believe Charles Darwin's theory of evolution is "definitely true," with another quarter saying it is "probably true." Half of the 2,060 people questioned were either strongly opposed to the theory or confused about it.

The Rescuing Darwin survey, published to coincide with the 200th anniversary of Darwin's birth and the 150th anniversary of the publication of "On the Origin of Species," found that around 10% of people chose young Earth creationism – the belief that God created the world sometime in the last 10,000 years – over evolution.

About 12% preferred intelligent design, the idea that evolution alone is not enough to explain the structures of living organisms. The remainder was unsure, often mixing evolution, intelligent design and creationism together. The survey was conducted by the polling agency ComRes on behalf of the Theos think tank.

James Williams, a lecturer at Sussex University, said, "Creationists ask if people believe in evolution. Evolution is a theory and a fact. You accept it because of the evidence. What the creationists have done is put a cloak of pseudo-science to wrap up their religious belief."

Later this month scientists and academics from across Europe will meet in Dortmund, Germany, to discuss evolution and creationism. It will be the first European conference of its kind to deal with different aspects of attitudes and knowledge related to evolution. They will discuss specific difficulties regarding the acceptance of evolution theory in their home countries.

Williams, who will give a paper presenting a British perspective on evolution and creationism in school science, said, "Evolution is very badly taught in schools so the results of the survey don't surprise me. On the other hand, creationism has traditionally been an issue in North America and there is a big problem in Australia and Turkey. It matters if people don't understand how science works." (End of article)

*****

How do you like that? Atheists just don't get it. They don't agree that people have rightly figured out that it takes *too much* "faith" to believe in evolution. Oh no! Rather it is that people are not bright enough to *understand* how "science" works. Or they are badly taught. After all, Creation is a North America—spelled, *United States,* issue. Americans are to blame again, I suppose. Our Christian roots influence the rest of the *intelligent* world improperly. Naturally this is seen as a "big problem" when we influence other nations to believe in Creation, primarily because to believe in Creation demonstrates a belief in God. Can't have that!

I don't think it is a big problem at all. Believing in the truth is never the problem. Believing in a lie is a far bigger one. And promoting a lie is even worse.

I will put my faith in the God who has proved Himself to mankind and to me over and over again. I have just barely scratched the surface of the available evidence that points to the God of the Bible being just who and what He says He is; An amazing God who loves me and who is preparing a place for me to live with Him for eternity. Now that is faith with a sure hope. My fervent prayer is that everyone will come to know the God I know and love.

God's desire for us to enjoy His love is well expressed in Eph. 3:17-19, "That Christ may dwell in your hearts by faith; that ye, being rooted and grounded in love, may be able to comprehend with all saints what is the breadth, and length, and depth, and height; and to know the love of Christ, which passeth knowledge, that ye might be filled with all the fullness of God."

Couple this with John 3:16, "For God so loved the world, that He gave His only begotten Son, that whosoever believeth in Him should not perish, but have everlasting life."

Surely, this hymn expresses the essence of God's love.

"Could we with ink the ocean fill
And were the skies
Of parchment made,
Were every stalk on earth a quill
And every man a scribe by trade,
To write the love of God above
Would drain the ocean dry,
Nor could the scroll
Contain the whole
Though stretched from sky to sky."

Beloved friend, you will have to fight your way to Hell over His love, with self-induced delusions. Please let His love melt your heart, and come to the bloody cross to Jesus.

About the Author

Biographical sketch of Rev. Floyd C. McElveen, missionary, evangelist, pastor, author.

"Mac" and his wife, Virginia

McElveen grew up in Mississippi, was religious but lost, nearly joined a cult, but was saved in La Grande, Oregon, on Sept. 29, 1949. He is married to a Mississippi beauty, Virginia McElveen, and is the father of three sons and one daughter.

"Mac" coached winning sports teams and taught very successfully in public school for 4 years, then went to Alaska as a missionary for 14 years, pioneering in evangelism and starting churches. His family lived largely on moose, caribou and bear meat and Alaskan salmon.

He built one church with logs wrested from the mountainside. McElveen started churches in Alaska, Idaho and Washington. He was appointed as the National Evangelist for the CBHMS (currently the "Mission Doors"), in 1980.

By God's good grace, McElveen saw God save thousands as he presented Jesus and His shed blood and risen life as missionary/evangelist/pastor all over the U.S.

"Mac" taught soul-winning classes and gave missionary challenges in scores, perhaps hundreds of churches, with thousands responding for soul-winning; full-time service and other decisions for Christ.

~ 106 ~
Big Mac Publishers

Other Books by Floyd C. McElveen

McElveen has over 1.1 million books in print, with 360,000 copies in Russia of "Evidence You Never Knew Existed" that is also published in Russia as "The Compelling Christ," but most often under the title of "Facts You Need to Know About."

This book is published by the thousands in China in Simplified Chinese and Mandarin, and in Korea. It has also been translated and published in Romania. Pastors are using the English version for outreach and discipleship, ordering and distributing hundreds of copies of this small but powerful book in their communities and to their people for outreach.

With the Bible League projects now complete, McElveen's books are in 8 or 9 languages. He recently wrote a book, published by Huntington Press, "The Disney Boycott." It is clear, concise, and gets to the heart of the real issue regarding the homosexual lifestyle being promoted by Disney.

McElveen's latest book published by Multnomah Publications is perhaps the crowning jewel. "Unashamed, A Burning Passion to Share the Gospel," has been endorsed by Dr. John Ankerberg, Dr. John Morris, the late Jerry Falwell and others.

Other books McElveen has written include, "The Beautiful Side Of Death," "The Mormon Illusion," currently published by Kregel, and "God's Word, Final, Infallible and Forever," published by Gospel Truth Ministries, and "The Call of Alaska," published by Promise Publishers. This book was subsidy-published, financed by Rocky McElveen, one of my sons.

For those interested in "Evidence You Never Knew Existed," the book can be ordered individually or in bulk from Gospel Truth Ministries, Fax 616-451-8907, or Ph. 616-451-4562, or Gospel Truth Ministries, 1340 Monroe Avenue, N.W., Grand Rapids, Michigan 49505.

Recently, McElveen has written a book, and acted as producer of "Jesus Christ-Joseph Smith, a Search for the Truth," in a DVD, with over 500,000 now distributed in English and thousands more translated into Spanish. The DVD is based on the book, so both have the same title.

Any effort not glorifying God is futile, but we pray and believe God has led us, in "Faith of an Atheist."

With my son Greg's hard work and editing, this book and one more will soon be on the market "So Send I You."

If this book has been a blessing to you, encouraged you or helped you to receive Christ, please share with us. If you have questions, please contact Mac or Virginia McElveen, by email: mac4christ@comcast.net or snail mail: 16 Sweet Bay Trail, Petal, Ms. 39465 or phone: 1-601-584-7123.

Dedication

I specifically dedicate this book to my beautiful wife of 60 years, Virginia McElveen, who for the first time in all the books I've written, lovingly wrote a chapter for me, "Mr., Where Does This Train Go?"

I also dedicate this book to Richard Hughes who helped me financially with the publishing expenses to get the book off the ground.

Floyd C. McElveen

Big Mac Publishers

www.ingramcontent.com/pod-product-compliance
Lightning Source LLC
Chambersburg PA
CBHW021201020426
42331CB00003B/162